THE INTERNATIONAL SERIES OF
MONOGRAPHS ON COMPUTER SCIENCE

General Editors

John E. Hopcroft Gordon D. Plotkin
Jacob T. Schwartz Dana S. Scott
Jean Vuillemin

THE INTERNATIONAL SERIES OF MONOGRAPHS ON COMPUTER SCIENCE

---

1. *The design and analysis of coalesced hashing*
   Jeffrey S. Vitter and Wen-Chin Chen
2. *Initial computability, algebraic specifications, and partial algebras*
   Horst Reichel

Initial Computability, Algebraic Specifications, and Partial Algebras

HORST REICHEL

Department of Mathematics and Physics
Magdeburg Technical University, GDR

CLARENDON PRESS · OXFORD

1987

Oxford University Press, Walton Street, Oxford OX2 6DP
Oxford New York Toronto
Delhi Bombay Calcutta Madras Karachi
Petaling Jaya Singapore Hong Kong Tokyo
Nairobi Dar es Salaam Cape Town
Melbourne Auckland
and associated companies in
Beirut Berlin Ibadan Nicosia

Oxford is a trade mark of Oxford University Press

Published in the United States
by Oxford University Press, New York

This book is copublished with Akademie-Verlag Berlin.
This edition is for sale throughout the world excluding the socialist countries

British Library Cataloguing in Publication Data
Reichel, Horst
Initial computability, algebraic specifications, and partial algebras. —
(The International series of monographs on computer science; 2)
1. Electronic digital computers — Programming
2. Electronic data processing — Mathematics
3. Algebra
I. Title II. Series
005.13'1 QA 76.6

ISBN 0-19-853806-5

Library of Congress Cataloging in Publication Data
Reichel, Horst
Initial computability, algebraic specifications, and partial algebras.
(The International series of monographs on computer science; 2)
Bibliography: p.
Includes index.
1. Electronic data processing — Mathematics.
2. Data structures (Computer science)
3. Programming languages (Electronic computers) — Semantics.
4. Partial algebras. I. Title. II. Series.
QA76.9.M35R45 1987 512 86-17952

ISBN 0-19-853806-5

Printed in the German Democratic Republic

Preface

Our purpose in writing this book is to prove that partial algebras form a powerful and suitable mathematical basis for specifying abstract programs on abstract data types. We hope to achieve this aim by the development of a special kind of partiality, called 'equational partiality', and by proving that structural induction on equationally partial algebras yields a sound and complete concept of computability. Additionally we show that this kind of partiality is one of some possible generalizations of the universal algebra of total algebras to partial ones.

The book presents both the developed theory of equational partiality, see Chapters 2 and 3, and its applications in specifications of abstract data types, see Chapters 1, 4, and 5. In addition to the material covered in the first edition, entitled *Structural Induction on Partial Algebras*, this edition deals with observability problems of abstract data types, i.e. the initial semantics of Chapters 1 and 4 is extended in the additional Chapter 5 to behavioural semantics.

Readers interested in the relationship of equational partiality to the general theory of partial algebras should consult BURMEISTER, *A Model Theoretic Oriented Approach to Partial Algebras*.

It is assumed that the reader is familiar with elementary set theory, with some universal algebra, especially with the concept of free algebras, and with the purposes of data abstraction in computer science. Some knowledge of elementary category theory could be helpful.

The content of the Chapters 1, 2, 3, and 4 is the result of long joint work with KLAUS BENECKE, ULRICH L. HUPBACH, and HEINZ KAPHENGST, and I am very thankful to these colleagues for that collaboration.

I would also like to thank ROD BURSTALL and DON SANNELLA for helpful discussions and suggestions during my visting fellowship at Edinburgh University, made possible by a research grant of SERC, where some parts of the book have been written. Special thanks are due to DON SANNELLA for his very careful checking of the manuscript and for many interesting suggestions.

Finally, I want to thank Oxford University Press for their helpfulness in preparing this book.

Magdeburg, March 1987 HORST REICHEL

Contents

Introduction

The main purpose of the book is to initiate an increasing use of partial algebras in computer science. There are many reasons for the inadequate state of application of partial algebras. Probably the bad reputation of the theory of partial algebras itself is the main cause.

The theory of partial algebras was initiated for mathematical reasons more or less independently by J. Schmidt and his group and by J. Słominski. In the years from 1965 to 1973 many important results were contributed by P. Burmeister, R. Kerkhoff, H. Hoeft and V. S. Poythress. But this first and purely mathematical impulse did not initiate a continously increasing development of the theory of partial algebras. The first results showed that a straightforward generalization of the theory of total algebras was not to be expected. Many different generalizations of the concept of homomorphism, of substructure, and of validity of term equations were introduced, and it was not quite clear which of them would be right ones. At this time there were not sufficiently many examples of partial algebras by means of which one could decide these questions.

Meanwhile the extensively developed theory of small categories and to a greater extend the application of universal algebra in computer science act as motives for the development of the theory of partial algebras. We present a special theory of partial algebras mainly oriented towards satisfying the requirements of applications of universal algebra in computer science and of the theory of small categories. The variety of possible generalizations of the theory of total algebras is thoroughly discussed in the book *A Model Theoretic Oriented Approach to Partial Algebras* by P. Burmeister.

Next we try to explain the basic idea of the theory of *equational partiality* presented. One reason for the frequent application of universal algebra in computer science is the use of structural induction on free algebraic structures. Very often one has to define a mapping $f\colon M \to P$ in a finite manner, where M is an infinite but finitely represented set. A typical example of this kind of problems is the definition of the semantics of a programming language. In this case the set M consists of all syntactically correct programs and P is a set of state transformations of an abstract interpreting machine. The finite representation of M is given by a phrase structure grammar. Let us consider the simple phrase structure grammar

$$\{A \leftarrow a, A \leftarrow bA\}$$

consisting of one non-terminal A and two terminals a, b, producing the infinite formal language

$$M = \{b^n a \mid n = 0, 1, 2, \ldots\}$$

The inductive structure of M is given by one nullary operation, corresponding to the production rule $A \leftarrow a$, and by one unary operation, corresponding to the production rule $A \leftarrow bA$. This set of associated operations, in algebraic terms the associated signature, is sometimes called the *abstract syntax*, because it abstracts from the spelling which is also defined by a phrase structure grammar. In this way we obtain an algebra $\mathbb{M} = (M; a, \omega)$ where $a \in M$ is the distinguished element and $\omega(b^n a) = b^{n+1} a$ for each $b^n a \in M$.

An application of structural induction to define a function $f: M \to P$ may be sketched in the following way: at first one has to find a finite algebraic theory T in such a way that M becomes the carrier of an initial T-algebra $\mathbb{M}$, where a T-algebra $\mathbb{A}$ is said to be initial if for every T-algebra $\mathbb{B}$ exactly one homomorphism $f: \mathbb{A} \to \mathbb{B}$ exists. In the second step of the inductive definition of $f: M \to P$ we have to define on the set P any T-algebra $\mathbb{P}$ so that the desired function is given by the uniquely determined homomorphism $f: \mathbb{M} \to \mathbb{P}$.

Intuitively a finite representation of an infinite set M by means of a finite algebraic theory T can be understood as follows: A finite algebraic theory is given by a finite set Ω of operations and by a finite set $\mathfrak{A}$ of axioms describing which combinations of the fundamental operations produce the same value. Among the fundamental operations the nullary operations have the function of atomic generating elements of M, so that any element of M can be represented as the value of a finite combination of non-nullary operations applied to atomic generating elements. If any two such combinations produce the same value, then this must be the consequence of the axioms. In the simplest case the theory T can be chosen in such a way that the set of axioms is empty. If we take the simplest non-trivial case, given by one atomic generating element, by one unary fundamental operation, and by the empty set of axioms, we get exactly Peano's axiomatization of natural numbers. Therefore, mathematical induction on natural numbers is a special case of structural induction.

In the case of formal languages any production rule of a context-free grammar, containing no non-terminal in its right-hand side, can be identified with a constant operation, or an atomic generating element, and any other production rule can be interpreted as a generating operation. The existence of different non-terminals corresponds in algebraic terms to different kinds of carrier sets, e.g. we have to use many-sorted or heterogeneous algebras. Combinations of generating operations corresponds to derivation trees. If the grammar is unambiguous the semantic meaning can be defined by structural induction. More detail in this direction can be found in the paper of Knuth (1968) or in the paper of Letitshevski (1968) and in the paper *Initial Algebraic Semantics* of the ADJ group (1975). However, the syntactic structure of existing programming languages cannot be completely described in an adequate way by context-free grammars. In general, most production rules are restricted by so-called context conditions. Algebraically we can interpret a production rule with a context condition as a partial operation with a necessary and sufficient domain condition. Another way to interprete production rules with context conditions is using error algebras as suggested by J. A. GOGUEN.

The same phenomenon of a necessary and sufficient domain condition of a partial operation can be observed in small categories. Small categories have exactly one partial operation, namely the partial composition of morphisms. And again, the partial binary operation is equipped with the necessary and sufficient domain condition in such a way that the product $x \circ y$ exists if and only if the target object of x is equal to the source object of y.

According to these observations in the theory of equationally partial algebras we equip each operator with a necessary and sufficient domain condition, given by a finite set of equations. A total operator is then an operator equipped with the trivial domain condition, for instance with the empty set of equations. We shall develop the theory of equationally partial structures in such a way that the notions and results of total algebras will be reproduced if every operator has a trivial domain condition. We shall prove that a certain syntactic hierarchy condition for the domain conditions of a signature (or of heterogeneous operator schemes) is necessary and sufficient for a nice and simple theory of equational partiality. With this approach we can grasp the inductive power of context-restricted production rules of real programming languages.

The application of structural induction on partial algebras will be demonstrated in Chapter 1 by specifications of several abstract data types and of parameterized abstract data types containing partial operations. We present these applications of structural induction on partial algebras without any preceding theoretical explanation in order to show that those specifications are not significantly harder then specifications using total operations only.

In the following three chapters we develop thoroughly the algebraic foundations of structural induction, of parameterized abstract data types with partial operations, and of functional enrichments of parameterized abstract data types.

In Chapter 2 we develop the algebraic theory of equational partiality. The basic notions are hierarchical equationally partial signatures (hep-signatures), hep-signatures with elementary implications as axioms, and finally hep-quasi-varieties as classes of models of hep-signatures with elementary implications. It will be shown that due to the hierarchy condition of the system of domain conditions the weakest concept of homomorphism between partial algebras is sufficient for the development of a sound theory of equational partial algebras that, for instance, completely reflects the structural behaviour of small categories. In this respect Chapter 2 mainly contributes to the universal algebra of partial algebras.

Chapter 3 develops the algebraic foundations of structural induction on partial algebras. It is mainly based on the existence of partial algebras that are freely generated by sets of generators and sets of defining relations on the generating elements. This concept, well-known in the theory of groups, is the algebraic key for a powerful and flexible approach to structural induction which works as well for total algebras as for partial algebras. This concept can also be used for constructions of partial quotient algebras. The theorem of homomorphisms in Section 3.4 represents the most significant difference in the structural behaviour of total algebras and of equationally partial algebras.

In Chapter 4 we deal with classes of equationally partial algebras that are exclusively motivated by semantic problems of parameterized abstract data types and by functional enrichments of parameterized abstract data types. These model classes are defined by so-called *canons*, i.e. by initially restricting algebraic theories. On the basis of this concept we define computable enrichments and definable enrichments of canons and prove some necessary conditions for definable and for computable enrichments of canons. In the last section we show how canons and canon morphisms can be used for the definition of the semantics of constructions that compose complex specifications from simpler ones. Finally we consider constructions like instantiations of parameterized abstract data types that preserve computable enrichments of canons.

Chapter 5 is devoted to the well-known fact that in general the formal representation of abstraction by isomorphism classes does not conform with real intuition in data abstraction. This has led to generalizations of the concepts of observability, distinguishability, and behaviour, known in the theory of abstract automata, to abstract data types. We will deal with an observability concept which is based on the partition of the set of sorts of a signature into visible and hidden sorts, and which reconstitutes the semantics of abstract data types as defined in the previous chapters if every sort is a visible one. Following this line we extend the concept of initial canons to behavioural canons and study computable enrichments of behavioural canons. We will see that initial computability as introduced in this book allows only computable enrichments specifying additional output functions.

1. Specifications of abstract data types with partial operations

1.1. Abstract data Types — ADTs

In connection with investigations of modularity concepts of large software projects the understanding of data types concepts was essentially modified. Whereas in programming languages like ALGOL 60, PL/1, and PASCAL a data type is considered to be a set, at least since the early seventies it seems to be generally accepted that data types are not just sets, but sets equipped with operations. For instance, if we deal with (homogeneous) *lists* the following operations can be considered to be relevant:

$$\left.\begin{array}{l}\text{equal(List,List)} \to \text{Bool},\\ \text{cons(Item,List)} \to \text{List},\\ \text{append(List,List)} \to \text{List},\\ \text{member(Item,List)} \to \text{Bool},\\ \text{empty} \to \text{List},\\ \text{length(List)} \to \text{Nat}.\end{array}\right\} \qquad (*)$$

Here we see that four different kinds of objects occur which are characterized by the type names 'Bool', 'List', 'Item', and 'Nat', respectively. This situation can be algebraically represented by a 4-sorted algebra A consisting of four carrier sets, namely
$A_{\text{Bool}} = \{T, F\}$, $A_{\text{Nat}} = \mathbb{N}$ (set of natural numbers), $A_{\text{Item}} = \{a, b, c, \ldots, y, z\}$ and A_{List} = set of all finite sequences over A_{Item}, and which consists of six fundamental operations:

$$\begin{array}{l}\text{equal}^A\colon A_{\text{List}} \times A_{\text{List}} \to A_{\text{Bool}},\\ \text{cons}^A\colon A_{\text{Item}} \times A_{\text{List}} \to A_{\text{List}},\\ \text{append}^A\colon A_{\text{List}} \times A_{\text{List}} \to A_{\text{List}},\\ \text{member}^A\colon A_{\text{Item}} \times A_{\text{List}} \to A_{\text{Bool}},\\ \text{empty}^A\colon \{\emptyset\} \to A_{\text{List}},\\ \text{length}^A\colon A_{\text{List}} \to A_{\text{Nat}}.\end{array}$$

Depending on the choice of the actual items, i.e. depending on the interpretation of the type name 'Item', we would get different concrete data types.

The example demonstrates that a concrete data type can be identified with a many-sorted algebra. But this understanding is not sufficient to put into practice the suggestion of Hoare (1972): "In the development of programs by stepwise refinement the programmer is encouraged to postpone the decision on representation of his data until after he has designed his algorithm and has expressed it as an abstract program operating on abstract data." The problem consists in finding a precise definition of an abstract data type and of an abstract program operating on an abstract data type.

The fundamental idea in abstract data type specifications is the mathematical representation of an abstract data type by a class of behaviourally equivalent many-sorted algebras. In the literature on abstract data types there are several different definitions of behavioural equivalence. At first we shall use the mathematically most simple version of equivalence which considers many-sorted algebras as behaviourally equivalent if they are isomorphic. In Chapter 5 we shall deal with a broader concept of behavioural equivalence, and in the last two chapters we shall make precise the notion of an abstract program operating on an abstract data type.

However, there are many very different interpretations of the preceding scheme (∗) which have nothing in common with the intended model of lists on a set of items. Although schemes of the form (∗) are used in the programming language ADA as the specification part of a package, they have no semantics. In ADA the semantics of a package is given by the implementation, or by the body of the package. This approach does not represent the behaviour of an abstract data type by the description of the properties of the available manipulation functions, but by giving a representative of the behavioural class defined in terms of other ADA-packages and of the basic types and basic functions of ADA. A disadvantage of this approach can be illustrated by the following example. If the natural numbers were represented by the particular model of Roman numbers, i.e. by I, II, III, IV, ..., anyone can see the difficulties in discovering the distributivity law of addition and multiplication of natural numbers. Therefore we prefer axiomatic tools for specifying the external behaviour of abstract data types.

Since our aim is the demonstration of the usefulness and feasibility of structural induction on partial algebras we will not explain the fundamentals of algebraic specifications of abstract data types. Readers not familiar with the basic ideas should consult the paper *Introduction to Data Algebra* by Zilles (1980) or *A Look at Algebraic Specifications* by the same author (see Zilles (1981)).

Coming back to the preceding example one can easily check that the intended interpretation satisfies the following properties expressed by the equations:

$$\left.\begin{aligned}
&\mathrm{length}(\mathrm{empty}) = 0,\\
&\mathrm{length}(\mathrm{cons}(x,w)) = \mathrm{length}(w) + 1,\\
&\mathrm{append}(\mathrm{empty},w) = w,\\
&\mathrm{append}(\mathrm{cons}(x,v),w) = \mathrm{cons}(x,\mathrm{append}(v,w)),\\
&\mathrm{equal}(w,w) = \mathrm{true},\\
&\mathrm{equal}(w,v) = \mathrm{equal}(v,w),\\
&\mathrm{equal}(w,\mathrm{append}(\mathrm{cons}(x,v),w) = \mathrm{false},\\
&\mathrm{equal}(\mathrm{cons}(x,w),\mathrm{cons}(y,v)) = \mathrm{equal}(w,v) \text{ and } x = y,\\
&\mathrm{member}(x,\mathrm{empty}) = \mathrm{false},\\
&\mathrm{member}(x,\mathrm{cons}(y,w)) = x{=}y \text{ or } \mathrm{member}(x,w),
\end{aligned}\right\} \qquad (**)$$

where w, v are variables for lists and x is a variable for items.

The next interpretation of the schema (*) shows that the system (**) of equations is not strong enough to exclude all those interpretations which are not isomorphic to the intended interpretation. Let the interpretation B be given by

$B_{\text{Bool}} = A_{\text{Bool}} = \{T, F\}$, $B_{\text{Nat}} = A_{\text{Nat}} = \mathbb{N}$,

$B_{\text{Item}} = A_{\text{Item}}$,

B_{List} = set of all functions $f\colon B_{\text{Item}} \to \mathbb{N}$ with $f(x) = 0$ for all but a finite number of items,

$\text{empty}^B = f_0$ with $f_0(x) = 0$ for all $x \in B_{\text{Item}}$,

$$\text{cons}^B(x, f) = f' \quad \text{with} \quad f'(y) = \begin{cases} f(y) & \text{for} \quad y \neq x \\ f(y) + 1 & \text{for} \quad y = x \end{cases}$$

for all $x, y \in B_{\text{Item}}$,

$\text{append}^B(f, g) = h$ with $h(x) = f(x) + g(x)$ for all $x \in B_{\text{Item}}$,

$$\text{member}^B(x, f) = \begin{cases} T & \text{iff } f(x) \neq 0 \\ F & \text{otherwise} \end{cases} \quad \text{for all} \quad x \in B_{\text{Item}},$$

$\text{length}^B(f) = \sum f(x)$ over all $x \in B_{\text{Item}}$,

$$\text{equal}^B(f, g) = \begin{cases} T & \text{iff } f(x) = g(x) \\ F & \text{otherwise.} \end{cases} \quad \text{for all} \quad x \in B_{\text{Item}}$$

The interpretation B represents multi-sets and does not represent lists. Evidently, A and B are not isomorphic since there is an equation which is satisfied by B but not by A, namely

$$\text{cons}(x,\text{cons}(y,w)) = \text{cons}(y,\text{cons}(x,w)).$$

This example indicates that we need an additional tool for excluding undesirable interpretations.

This additional exclusion of interpretations will be done by the concept of *initial satisfaction* of axioms. Roughly, we can say that an interpretation C of (*) initially satisfies the set (**) of equations if C satisfies only those equations which are consequences of (**), where an equation $t_1 = t_2$ is called a consequence of a set G of equations if every algebra D satisfying each equation in G also satisfies $t_1 = t_2$. The interpretation A shows that the equation

$$\text{cons}(x,\text{cons}(y,w)) = \text{cons}(y,\text{cons}(x,w))$$

is not a consequence of (**), so that B does not initially satisfy the set of equations given by (**).

Next we will make the concept of initial satisfaction more precise. Readers wanting a more detailed explanation should look into the paper *An Initial Algebra Approach to the Specification, Correctness, and Implementation of Abstract Data Types* of the ADJ group (see ADJ (1978)).

In terms of universal algebra one can say that an algebra C of scheme (*) initially satisfies (**), iff for every algebra D of scheme (*) satisfying (**) there is exactly one homomorphism $f: C \to D$. Obviously, any two algebras are isomorphic if they initially satisfy (**). This shows that initial satisfaction conforms to the specifications of isomorphism classes of algebras.

However, it was shown by BERGSTRA and MEYER (1983) that there are very simple abstract data types that are not specifiable by finite sets of equations. They show that the set of all finite subsets of natural numbers with the cardinality function may serve as such a counterexample. A finite specification of that data type requires conditional equations of the form

$$t_1 = r_1, \ldots, t_n = r_n \to t = r.$$

Next we introduce the syntactic form for writing specifications of abstract data types by means of a simplified version of list that differs slightly from the preceding version since we are not yet able to specify lists with an arbitrary set of items. In the following specification we use the fixed set $\{a, b, \ldots, y, z\}$. Additionally we omit the specification of the equality and of the membership relation.

LIST **is definition**
  sorts Nat, Item, List
  oprn zero → Nat
    succ(Nat) → Nat
    a,*b*, …, *y*,*z* → Item
    empty → List
    cons(Item,List) → List
    append(List,List) → List
    length(List) → Nat
  axioms *v*,*w*,: List, *x*: Item
    length(empty) = zero
    length(cons(*x*,*w*)) = succ(length(*x*))
    append(empty,*w*) = *w*
    append(cons(*x*,*v*),*w*) = cons(*x*,append(*v*,*w*))
end LIST

The key-word **definition** indicates initial satisfaction for the following axioms. After the key-word **sorts** the new sort names follow which correspond to the specified data types. After the key-word **oprn** the sequence of fundamental operators follows, where in brackets, following the operator name, the number, the ordering and the required sorts of the arguments of that operator are declared by an n-tuple of sort names. The sort name after the arrow declares the result sort. This notation is intentionally similar to function declarations in programming languages. This implies that in the preceding specification 'zero → Nat' denotes a constant of type 'Nat', i.e an operator without any arguments that produces an object of type 'Nat' and that

$$\text{cons(Item,List)} \rightarrow \text{List}$$

denotes an operator which needs two arguments, the first is of sort 'Item' and the second of sort 'List', and it denotes that the resulting value is again of sort 'List'.

$$a,b,\ldots,y,z \rightarrow \text{Item}$$

declares a set of constant operators with the names $a,b,\ldots,y,z$ all of type 'Item'. After the key-word **axioms** the declaration of the variables follows that are used in the following axioms. In the preceding specification each axiom is an equation.

A verification of the specification LIST would require the proof of the *initiality* of the interpretation which is suggested by the chosen names of sorts and of operators.

The next example illustrates the use of conditional equations by a specification of the set of all finite subsets of natural numbers equipped with the cardinality function.

CARDINALITY **is definition**
    sorts Nat, Sets, Bool
    oprn zero → Nat
        succ(Nat) → Nat
        true, false → Bool
        eq(Nat,Nat) → Bool
        $\emptyset \rightarrow$ Sets
        Sets $\cup$ {Nat} → Sets
        card(Sets) → Nat
        Nat $\in$ Sets → Bool
    axioms x,y: Nat, m: Sets
        $\text{eq}(x,x) = \text{true}$
        $\text{eq}(x,y) = \text{eq}(y,x)$
        $\text{eq}(\text{zero},\text{succ}(x)) = \text{false}$
        $\text{eq}(x,y) = \text{eq}(\text{succ}(x),\text{succ}(y))$
        $(m \cup \{x\}) \cup \{x\} = m \cup \{x\}$
        $(m \cup \{x\}) \cup \{y\} = (m \cup \{y\}) \cup \{x\}$
        $(x \in \emptyset) = \text{false}$
        $x \in (m \cup \{x\}) = \text{true}$
        if $(y \in m) = \text{false}$ **then** $y \in (m \cup \{x\}) = \text{eq}(x,y)$
        $\text{card}(\emptyset) = \text{zero}$
        if $(x \in m) = \text{false}$ **then** $\text{card}(m \cup \{x\}) = \text{succ}(\text{card}(m))$
end CARDINALITY

In order to improve the readability of specifications we also allow the functional notation of operator application both with *infix notation*, as it is done in the case 'Nat ∈ Sets → Bool', and the *mixed-fix notation*, which is used for the operator of adding a natural number to a finite set 'Sets ∪ {Nat} → Sets'.

A further step in this direction could be the abbreviation of an equation

$$(x \text{ rel } y) = \text{true} \quad \text{by} \quad x \text{ rel } y,$$

and of an equation

$$(x \text{ rel } y) = \text{false} \quad \text{by} \quad x\, \overline{\text{rel}}\, y,$$

where 'rel' is the name of a binary infix operator producing truth values. Using these conventions some of the preceding axioms could be written as follows:

$$x \notin \emptyset, \quad x \in (m \cup \{x\}),$$

$$\text{if} \quad y \notin m \quad \text{then} \quad y \in (m \cup \{x\}) = \text{eq}(x,y)$$

$$\text{if} \quad x \notin m \quad \text{then} \quad \text{card}(m \cup \{x\}) = \text{succ}(\text{card}(m)).$$

Assuming that the preceding examples sufficiently explain the way of writing specifications of abstract data types with total operations we turn now to specifications of data types with partial operations.

Without any algebraic foundation partial operations are extensively used in attributed grammars in the form of restricted production rules, where the restriction of production rules may be expressed by equations on attributes. By the following example we will demonstrate that the initial approach of the definition of syntax and semantics of context-free languages can be extended to languages generated by a grammar with restricted production rules. Also in this case syntax and semantics can be defined via homomorphisms from an initial partial algebra built up from derivation trees of the grammar.

To start with simpler examples we first specify the set of all those finite sequences of natural numbers that contain no natural number more than once. For an inductive construction of all such sequences the adding of a natural number n to a sequence r will be restricted to the case that n is not contained in r. This leads to a partial operation.

Since we need natural numbers for several specifications we start with the specification NAT which will be extended to EXAMPLE 1 afterwards.

NAT **is definition**
  sorts Nat, Bool
  oprn true, false → Bool
    zero → Nat
    succ(Nat) → Nat
    eq(Nat,Nat) → Bool
    and(Bool,Bool) → Bool
    or(Bool,Bool) → Bool

axioms x,y: Nat, z: Bool
eq(x,x) = true
eq(x,y) = eq(y,x)
eq(zero,succ(x)) = false
eq(succ(x),succ(y)) = eq(x,y)
and(true,true) = true
and(z,false) = and(false,z) = false
or(false,false) = false
or(z,true) = or(true,z) = true
end NAT

EXAMPLE 1 **is** NAT **with definition**
sorts Rseq
oprn nil → Rseq
Nat ∈ Rseq → Bool
add(x:Nat, r:Rseq **iff** $x \notin r$) → Rseq
axioms x,y: Nat, r: Rseq
$x \notin$ nil
if $x \notin r$ **then** ($y \in$ add(x,r)) = eq(x,y)
end EXAMPLE 1

In the specification EXAMPLE 1 the specification NAT is used as prerequisite. In this way complex specifications can be written by stepwise enrichments of presupposed specifications.

Of course, this example could also be specified with a total adding operator. In this case we would have to add a new axiom that guarantees that the addition of a natural number n contained in the sequence r does not change the sequence r, i.e.

if $x \in r$ **then** add(x,r) = r.

Another way of making the adding operator to a total one would be the use of a so-called error value, as it was suggested by GOGUEN (1977). The resulting specification is then

EXAMPLE 2 **is** NAT **with definition**
sorts Rseq
oprn nil → Rseq
error → Rseq
add(Nat,Rseq) → Rseq
axioms x,y: Nat, r: Rseq
$x \notin$ nil
$x \notin$ error
if $x \notin r$ **then** $y \in$ add(x,r) = eq(x,y)
add(x,error) = error
if $x \in r$ **then** add(x,r) = error
end EXAMPLE 2

It is evident that these three specifications do not specify one and the same abstract data type. It is also meaningless to say that only one of these specifications is the correct one. It depends on the context of application in which the abstract data type will be used. Only from this context one gets the information to decide wether or not a programmer should be forced to think about the applicability of the add-operator before using that operator, or if the programmer should provide exception handling which comes into account after an application of the add-operator, or if it is finally of no interest from outside if a programmer tries the addition of a natural number to a sequence that is already contained in the sequence. In the last case the occurence of such an addition would be detected by the abstract data type itself and according to the axiom

$$\textbf{if} \quad x \in r \quad \textbf{then} \quad \text{add}(x,r) = r$$

the abstract data type would act correctly.

This discussion shows that any specification language should give the designer of a software project the possibility to decide in which way partiality is considered. This is one reason why we claim that a specification language should allow the specification of partial operation. Furthermore it is proved in the theoretical considerations of Section 3.4. that there is no way to describe the structural behaviour of partial algebras in an adequate way in terms of total algebras. Thus, there are practical and theoretical reasons to include partial operations in specification languages.

In the next example we specify trees whose edges take values in the natural numbers in such a way that each path from the root to any leaf induces a decreasing sequence of edge values. This specification is divided into two parts. In the first part we do not specify any new type, but we enrich NAT by additional operations that are very useful for the following specification of *decreasing edge-valued trees.*

DECREASING EDGE-VALUED TREES is NAT **with definition**
    oprn max(Nat,Nat) $\rightarrow$ Nat
        Nat $<$ Nat $\rightarrow$ Bool
    axioms x,y: Nat
        $x >$ zero
        zero $<$ succ(**d**)
        $(x < y) = (\text{succ}(x) < \text{succ}(y))$
        $\max(x,x) = x$
        $\max(x,y) = \max(y,x)$
        if $x < y$ **then** $\max(x,y) = y$
    with definition
    sorts Tree, Treeseq

oprn nil $\rightarrow$ Treeseq
add(Treeseq, r:Tree, n:Nat **iff** t-attr(r) $< n$) $\rightarrow$ Treeseq
combine(Treeseq) $\rightarrow$ Tree
t-attr(Tree) $\rightarrow$ Nat
seq-attr(Treeseq) $\rightarrow$ Nat
axioms n: Nat, r: Tree, s,s_1,s_2: Treeseq
seq-attr(nil) = zero
t-attr(combine(s)) = seq-attr(s)
if s_1 = add(s_2,r,n) **then** seq-attr(s_1) = max(seq-attr(s_2),n)
end DECREASING EDGE-VALUED TREES

The functional enrichment of the specification NAT in the first part of the preceding specification needs no additional explanations. The simplest tree consisting of exactly one node and of no one edge is represented by the formal expression 'combine(nil)'. This tree will be pictured by a small circle. The objects of sort 'Treeseq' are special sequences of pairs consisting of a tree and a natural number. Both for trees and for objects of the sort 'Treeseq' respectively an attribute is defined which associates a natural number with each object of either sorts, respectively named 't-attr' and 'seq-attr'. The action of the combine operator can be pictorially described as follows:

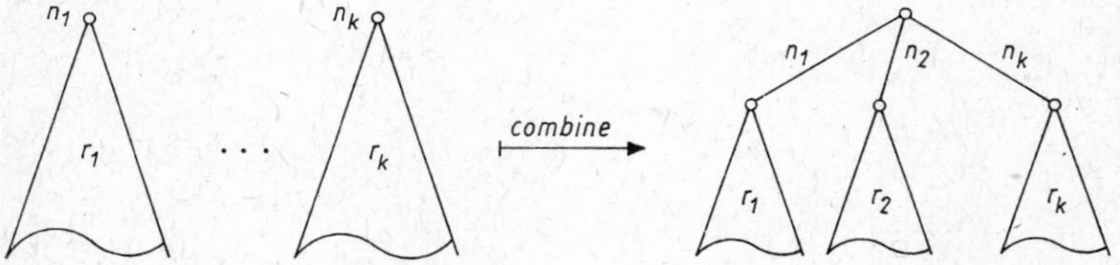

The following pictorially represented tree is not an object of sort 'Tree' since the leftmost path induces the sequence 3,5, which is not a decreasing sequence:

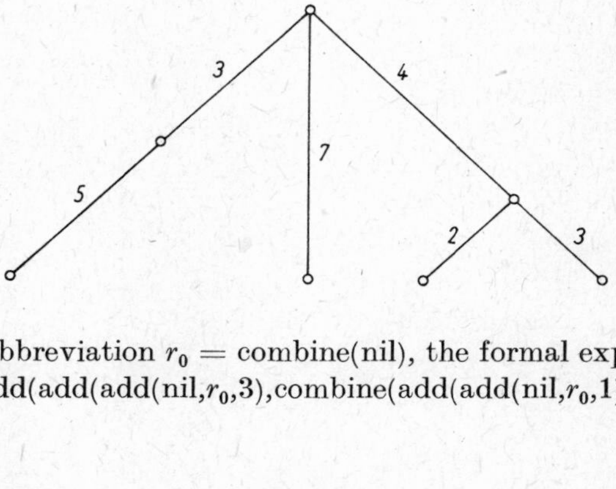

If we use the abbreviation r_0 = combine(nil), the formal expression
r = combine(add(add(add(nil,r_0,3),combine(add(add(nil,r_0,1),r_0,3)),4),r_0,2))

corresponds to the following pictorial representation:

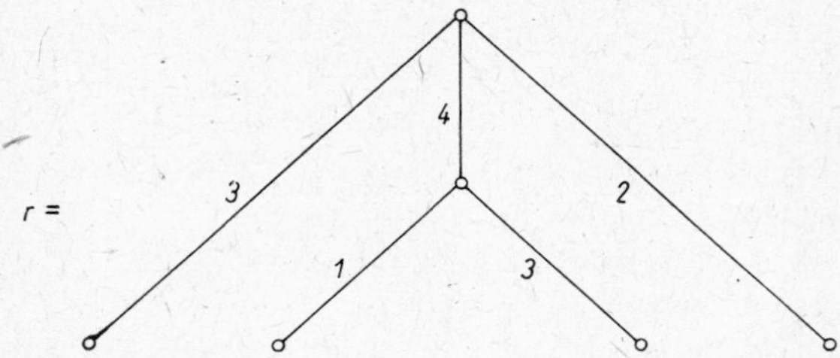

Furthermore the following equations hold:

$$\text{t-attr}(r) = 4$$

$$\text{t-attr}(\text{combine}(\text{add}(\text{add}(\text{nil},r_0,1),r_0,3))) = 3.$$

Now it seems necessary to say some words concerning partiality. We have associated with the add-operator a necessary and sufficient domain condition. The necessity of the domain condition restricts the class of all possible interpretations in the intended way. In the preceding example exactly the necessity implies that only decreasing sequences of edge values are generated by any path from the root to any leaf. In comparision to attribute grammars the necessity of the domain condition represents the context condition. However, if we made available only tools for restricting the applicability of operations, then the concept of initial partial algebras would not work in the expected way. For partial algebras initiality has an additional effect. Initiality causes minimal domains of partial operations. With respect to the specification of decreasing edge-valued trees this means that the domain of the add-operation would become empty if the domain condition were only necessary. In this case 'nil' would be the only object of sort 'Treeseq' and 'combine(nil)' would be the only object of sort 'Tree'. Now we see that the sufficiency of domain conditions of partial operators is the tool to overcome this weakness of ordinary initial partial algebras. Thus, necessary and sufficient domain conditions are a proper algebraic counterpart to context conditions in attribute grammars, because a production rule is applicable if and only if the context condition is satisfied.

Next we will informally discuss the validity of equations and of conditional equations for partial algebras. A mathematically precise discussion of this topic is given in Chapter 2.

The left-hand side and right-hand side of an equation is built up from variables and operators as formal expressions in the same way as expressions in programming languages. If the expressions of an equation do not contain partial operators, then there is no problem. An interpretation of the sort names and of the operators compatible with the declarations of the operators delivers an algebra A. A valuation of the variables of an equation in an algebra A is called a *solution* of the equation in A if the left-hand and right-hand expressions evaluate to the same value. It is worth mentioning that an evaluation must assign a value to each variable (i.e. variables may not represent 'undefined'). An equation is satisfied in an algebra A if each possible evaluation of the variables is a solution of that equation in A. If the left-hand expression or the right-hand expression contains one or more partial operators, then these expressions are not evaluable for each valuation of the variables. In this case an evaluation of the variables in a partial algebra A is said to be a solution in A if both sides of the equation are defined and if the resulting values are equal. Straightforwardly we define the set of solutions of a set of equations (all have to be defined over the same system of variables) as the intersection of the set of solutions of the single equations. Since a conditional equation is of the form

$$\textbf{if } G \textbf{ then } e_1 = e_2$$

where G is a finite, possibly empty set of equations, we say that a partial algebra A satisfies that conditional equation if each solution of G in A is a solution of $e_1 = e_2$ in A, too. For reasons of convenience we state by definition that any evaluation of a system of variables in a partial algebra A is a solution of the empty set of equations. Applying this to a conditional equation with an empty premise, i.e. applying it to

$$\textbf{if } \emptyset \textbf{ then } e_1 = e_2,$$

we obtain that a partial algebra satisfies that conditional equation if and only if both e_1 and e_2 are evaluable for each evaluation of the variables and if equal values are produced. According to this fact, we may consider equations as abbreviated notations of conditional equations with an empty premise. This implies that all axioms in a specification are conditional equations, but some of them may have a trivial (i.e. an empty) premise.

There is another effect deserving attention. Conditional equations cause *applicability requirements*. We illustrate this effect by the axiom

$$\textbf{if } x \notin r \textbf{ then } (y \in \text{add}(x,r)) = \text{eq}(x,y)$$

in EXAMPLE 1. The only partial operator in this specification is the add-operator. Bearing in mind the domain condition of the add-operator the premise says that add(x,r) is defined. If we substituted the premise $x \notin r$ by the empty set of equations, then the axiom

$$\textbf{if } \emptyset \textbf{ then } (y \notin \text{add}(x,r)) = \text{eq}(x,y)$$

would imply that the add-operator would become a total operator, and this would semantically contradict the intention of the domain condition. But this effect of conditional equations opens another style of specifications.

One can omit the domain condition in the declaration of the operator and can achieve its effect by two additional axioms expressing the sufficiency and the necessity of the domain condition, respectively. In this way we would obtain the following specification EXAMPLE1*, which is semantically equivalent to the previous specification EXAMPLE1.

> EXAMPLE1* **is** NAT **with definition**
>   **sorts** Rseq
>   **oprn** nil $\to$ Rseq
>   Nat $\in$ Rseq $\to$ Bool
>   add(Nat,Rseq) ⤏ Rseq
>   **axioms** x,y: Nat, r: Rseq
>   **if** $x \notin r$ **then** add(x,r) = add(x,r)
>   **if** add(x,r) = add(x,r) **then** $x \notin r$
> $x \notin$ nil
>   **if** $x \notin r$ **then** ($y \in$ add(x,r)) = eq(x,y)
> **end** EXAMPLE1*

Now the arrow ⤏ indicates that the operator 'add' may be interpreted by a partial operation where its domain is defined by the applicability requirements of the axioms and not by a domain condition as part of the declaration of the operator.

An arrow $\to$ indicates an operator where its domain is defined by a domain condition as part of the declaration of that operator. However, this does not in general prevent the axioms from causing stronger applicability requirements than the domain condition. In specifications with domain conditions one has always to consider the compatibility of domain conditions and axioms.

We do not prefer one of these styles of specifications. If the domain condition is simple we prefer the specification of explicit domain conditions within the declaration of the operator just as in EXAMPLE1. Unfortunately it is not a simple task to check whether the axioms cause additional applicability requirements. A corresponding condition will be discussed in Section 2.5. However, we will also see examples for which a specification in the style of EXAMPLE1* seems to be more convenient. The specification of the semantics of a while ... do ...-statement, given in the following section, is such an example.

In a specification of an ADT one should use minimal systems of operations in order to improve the readability. To achieve this simplicity one should specify only those operators that are necessary for the generation of all objects of the intended data type, and that are necessary for the domain conditions and for the specification of the intended interrelations of the generating operators. In this sense the specification of decreasing edge-valued trees is minimal. If we would apply this principle to natural numbers and truth values, we would introduce the constants 'zero', 'true', and 'false' and the unary successor operator. But there are different points of view of assessing specifications. The suggested purity of specifications improves comprehensibility, but it is evidently a disadvantage concerning the value in use. The improvement of the value in use of an ADT generally requires functional enrichments of basically specified ADTs.

According to the preceding considerations we discuss functional enrichments of ADTs in the following.

First of all, for each ADT the specification of the identity should be added. Therefore we start with the specification of the identity of restricted sequences of natural numbers and of the identity of decreasing edge-valued trees.

The main problem in functional enrichments of ADTs is the consistency and persistency of the new operators and axioms with the given ADT. This means that initial satisfaction of the enriched specification does not cause identification of existing objects of the given ADT, and that by means of the added operators no new objects will be created, and finally that the domains of given operators will not be changed.

The following enrichment of EXAMPLE1 does not fulfil the requirement of consistency and persistency:

> EXAMPLE 3 **is** EXAMPLE1 **with definition**
>   **oprn** r-eq(Rseq,Rseq) $\rightarrow$ Bool
>   **axioms** m: Rseq
>   r-eq(m,m) = true
> **end** EXAMPLE 3

Because of initiality the new operator 'r-eq' generates for each pair (m,m') of different restricted sequences a new object of sort 'Bool'. Thus, the given enrichment does not preserve the semantics of the given specification. The generation of new objects is a consequence of the absence of axioms that force r-eq(m,m') = false for each pair with different components. However, this statement of the intended property of the new operator 'r-eq' uses the very identity of restricted sequences which we want to specify. To break through this circle we go back to the inductive generation of restricted sequences and try to specify inductively the intended semantics of 'r-eq'.

> EXAMPLE 4 **is** EXAMPLE 1 **with definition**
>   **oprn** r-eq(Rseq,Rseq) $\rightarrow$ Bool
>   **axioms** x,y: Nat, m,m',m'': Rseq
>   r-eq(m,m) = true
>   **if** $x \notin m'$, $y \notin m''$ **then** r-eq(add(x,m'),add(g,m''))
>   = and(eq(x,y),r-eq(m',m''))
> **end** EXAMPLE 4

The specification EXAMPLE 4 seems to be an inductive specification of the intended equality. A more careful study points to the open problem of checking the equality of restricted sequences of different length. For instance, there is no way of proving by means of only the given two axioms that 'r-eq(nil,add(x,nil))' is equal to 'true' or to 'false'. The lack of a formal proof because of initiality, again necessitates the creation of new objects of sort 'Bool'. The failure in the preceding specification could be corrected by an additional specification of a length function and by adding the axiom that causes the inequality of restricted sequences of different length according to the following specification.

EXAMPLE 5 is EXAMPLE 4 with definition
  oprn length(Rseq) $\rightarrow$ Nat
  axioms x: Nat, m,m': Rseq
    length(nil) = zero
    if $x \notin r$ **then** length(add(x,r)) = succ(length(r))
    if eq(length(m),length(m')) = false
      then r-eq(m,m') = false **fi**
end EXAMPLE 5

EXAMPLE 5 is not aimed to give a most elegant specification of the equality of restricted sequences of natural numbers. The main object was the demonstration of problems that appear in connection with functional enrichments of ADTs. However, we have dealt only with the creation of new objects as a consequence of insufficient systems of axioms.

In the following we discuss the compatibility with domain conditions of given operators. If we remove the condition '$y \notin m''$' from the premise of the axiom that handles restricted sequences of equal length, then this modified axiom causes the totality of the add-operator. This axiom says that for all assignments of the variables x,y: Nat, m',m'': Rseq satisfying $x \notin m'$ the formal expression r-eq(add(x,m'),add(y,m'')) becomes evaluable, and so the subexpression add(y,m'') becomes evaluable too for all such assignments. But now there is no precondition for the addition of y to m'', and therefore the add-operator becomes total.

The foregoing discussion shows that the partiality of generating operators does not seriously complicate inductive specifications of functional enrichments. The only thing that has to be born in mind is that the premise has to contain all those conditions that guarantee the evaluability of all expressions and subexpressions of the conclusion.

The third type of incompatibility of an enrichment with a given specification concerns the identification of objects. We modify the same axiom as before, but now we substitute the truth function 'and' by 'or'. In this case the modified axiom would cause the identification of both truth values. By means of the modified axiom we would obtain for m' = add(z,nil),

$m'' =$ nil, and $x = y$ the equality

$$\text{true} = \text{or}(\text{eq}(x,x),\text{r-eq}(m',m'')) = \text{r-eq}(\text{add}(x,m'),\text{add}(x,m''))$$

and because of length(add(x,m')) $= 2 + 1 =$ length(add(x,m'')) the last axiom implies r-eq(add(x,m'),add(x,m'')) $=$ false.

Next we specify the equality of decreasing edge values trees without an extensive discussion of failures.

DECREASING EDGE-VALUED TREES WITH EQUALITY **is**
DECREASING EDGE-VALUED TREES
  with definition
  oprn t-eq(Tree,Tree) → Bool
    s-eq(Treeseq,Treeseq) → Bool
  axioms s,s': Treeseq, r,r': Tree, n,n': Nat
    t-eq(combine(s),combine(s')) = s-eq(s,s')
    s-eq(nil,nil) = true
    s-eq(s,s') = s-eq(s',s)
    if t-attr(r) $< n$ **then** s-eq(add(s,r,n),nil) = false
    if t-attr(r) $< n$, t-attr(r') $< n'$ **then**
      s-eq(add(s,r,n),add(s',r',n'))
      = and(and(eq(n,n'),t-eq(r,r')),s-eq(s,s')) **fi**
end DECREASING EDGE-VALUED TREES WITH EQUALITY

Even if in many papers on abstract data types it is assumed that for each sort the equality is available, the equality is not initially specifiable for each sort in each ADT. The undecidability of the word problem in semigroups and other examples of recursively undecidable problems allows the specification of ADTs where the equality is not recursively decidable, so that according to Section 4.2 the equality is also not initially specifiable. Later we will see that in an ADT with partial operations the initial specifiability of the equality is equivalent to an interesting concept, namely to the recursiveness of that ADT. Informally we call an ADT recursive if the isomorphism class, which semantically represents the ADT, contains an algebra with recursively enumerable carrier sets and with partial recursive fundamental operations, i.e. if there is a computable *prototype* of the given ADT. Unfortunately it is recursively undecidable whether an arbitrary specification has a computable prototype or not.

We will conclude this section with two further enrichments of DECREASING EDGE-VALUED TREES. By the first functional enrichment we specify a function that associates with an object t of sort 'Tree' a natural number depth(t) that gives the maximal length of all paths over t from the root to leaves. By the second functional enrichment we specify a function that gives for an object of sort 'Tree' the sum over all edge values. For the following pictorial representation of a decreasing edge-valued tree t_0, both depth(t_0) $= 3$ and sum(t_0) $= 18$ hold.

$t_0 =$ 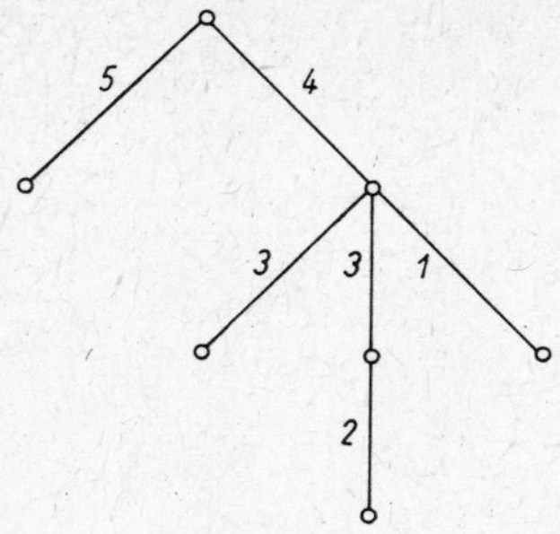

EXAMPLE 6 **is** EDGE-VALUED TREES **with definition**
oprn depth(Tree) $\rightarrow$ Nat
s-depth(Treeseq) $\rightarrow$ Nat
axioms n: Nat, r: Tree, s: Treeseq
depth(combine(nil)) $=$ zero
s-depth(nil) $=$ zero
if t-attr(r) $< n$ **then**
depth(combine(add(s,f,n))) $=$ succ(s-depth(s)) **fi**
if t-attr(r) $< n$ **then**
s-depth(add(s,r,n)) $=$ max(s-depth(s),depth(r)) **fi**
end EXAMPLE 6

This style of defining a function over an ADT seems to be very natural. We have only to think about computing the intended value if we know the corresponding value of all constituents with respect to all generating operators and how to compute the intended value for basic generating objects. Since not all constituents need be of the same sort it may happen that a simultaneous specification of an auxiliary operator is very helpful if not actually necessary. In the preceding specification the operator 's-depth' serves as such an auxiliary function. In the same manner we can specify the second function.

EXAMPLE 7 **is** DECREASING EDGE-VALUED TREES **with definition**
oprn Nat $+$ Nat $\rightarrow$ Nat
sum(Tree) $\rightarrow$ Nat
s-sum(Treeseq) $\rightarrow$ Nat
axioms x,y: Nat, r: Tree, s: Treeseq
zero $+ y = y$
succ(x) $+ y =$ succ($x + y$)
sum(combine(s)) $=$ s-sum(s)
s-sum(nil) $=$ zero
if t-attr(r) $< x$ **then**
s-sum(add(s,r,x)) $=$ (s-sum(s) $+$ sum(r)) $+ x$ **fi**
end EXAMPLE 7

Finally we point out that the demonstrated style of defining a function on an ADT significantly differs from the imperative style as used in ordinary programming languages. In terms of an imperative programming language it would not be possible to define one of the functions depth(Tree) → Nat or sum(Tree) → Nat using only the functions made available by the specification DECREASING EDGE-VALUED TREES because of the lack of operators which decompose compound objects. In functional or logical programming languages like Standard ML, HOPE or PROLOG the situation is the same as would apply here if we restricted our attention to total operations.

1.2. Parameterized abstract data types — PADTs

The examples of the previous section are so simple that formal specifications are not necessary for dealing with such objects. Formal specifications are destined for complex problems for which no suitable mathematical models exist. However, it is an illusion to expect that specifications of complex problems have to be simple. First experiences with formal specifications show that formal specifications may also become complex and therefore hard to read, hard to write, and hard to modify as a whole.

The main advantage of formal specification methods is the possibility of breaking down the complex problem into well-defined pieces of understandable size. It was the specification language CLEAR, invented by Burstall and Goguen (1977) that focused interest on facilities for writing *structured specifications*. By means of parameterized specifications a large real-world problem can be broken down and built up from small, easy to understand reusable pieces.

The problem of structuring large specifications is one of the central research problems in program methodology. A detailed discussion of structuring principles or a review of existing approaches to structures of specifications is beyond the range and aim of this book. In the opinion of the author there are at least two concepts that should be generally accepted, namely the CAT-concept of Goguen and Burstall (1980) and the idea of *liberal institutions* of Burstall and Goguen (1983).

The CAT-concept reflects the important fact that structuring of specification is only one dimension of the problem. Formal of specification methods should not only be used for specifications of the functional behaviour of a large program before writing that program in a programming language. If the *conceptual distance* between the basic notions of the specification and the basic notions of the programming language is very large, intermediate conceptual levels are necessary. Formal specifications should also be used for descriptions of transformations of the basic notions of one conceptual level into compound notions of the next lower conceptual level. The composition of specification that represents transformations in the direction from the highest conceptual level to the conceptual level of the used programming language gives the second dimension of complexity of large programs. In this way two kinds of compositions of specifications exists. GOGUEN and BURSTALL call these composition the *horizontal* and the *vertical composition*.

Liberal institutions are aimed to give a mathematical foundation that allows the definition and investigation of specification-building operations largely independent of the concrete mathematical semantics that is used for the specifications of ADT. In the terminology of liberal institutions it is the aim of this book to give an introduction, an investigation and an illustration of two special liberal institutions which will be called *initial canons* and *behavioural canons,* according to the report of HUPBACH, KAPHENGST and REICHEL (1980). As mentioned earlier the main characteristic of both canons is the availability of partial operations.

Before we generally explain the facilities for textual representations of structured initial canons we will give a simple example of a parameterized abstract data type. We will specify the PADT of finite directed paths over directed graphs. As distinct from specifications of ADTs we have now not only to specify an isomorphism class of many-sorted algebras. In the following example we have at first to specify the class of directed graphs which act as parameters. On the basis of the specification of the parameter class we have to specify the uniform construction of the set of directed paths over an arbitrary directed graph.

DIRECTED GRAPH **is requirement**
  sorts Edges, Nodes
  oprn begin(Edges) → Nodes
    end(Edges) → Nodes
end DIRECTED GRAPH

The key word **requirement** indicates that interpretation of the following sort names and of the following operators are not restricted to initial satisfaction. Since DIRECTED GRAPH does not contain any axiom, each two-sorted algebra

$$G = (G_{\text{Edges}}, G_{\text{Nodes}}; \text{begin}^G\colon G_{\text{Edges}} \to G_{\text{Nodes}}, \text{end}^G\colon G_{\text{Edges}} \to G_{\text{Nodes}})$$

is a model of the specification DIRECTED GRAPH, where G_{Edges} and G_{Nodes} are the sets of edges and nodes of G, respectively, and an edge $x \in G_{\text{Edges}}$ is directed from $\text{begin}^G(x) \in G_{\text{Nodes}}$ to $\text{end}^G(x) \in G_{\text{Nodes}}$.

By the next step we specify the parameterized construction of the set of finite directed paths.

FINITE PATHS **is** DIRECTED GRAPH **with definition**
  sorts Path
  oprn nil(Nodes) → Path
    add(w:Path,x:Edges **iff** p-end(w)=begin(x)) → Path
    p-end(Path) → Nodes

```
    axioms e: Nodes, w1,w2: Path, x: Edges
        p-end(nil(e)) = e
        if w2 = add(w1,x) then p-end(w2) = end(x)
end FINITE PATHS
```

If we want to extend an interpretation G of DIRECTED GRAPH to an interpretation $W(G)$ of the extended initial canon FINITE PATHS the key word **definition** requires that $W(G)$ is an *initial extension* of G. This means

(i) The given interpretation G has to be preserved by the initial extension $W(G)$, i.e.

$$W(G)_{\text{Edges}} = G_{\text{Edges}}, \quad W(G)_{\text{Nodes}} = G_{\text{Nodes}},$$

$$\text{begin}^{W(G)} = \text{begin}^G, \quad \text{end}^{W(G)} = \text{end}^G.$$

(ii) The interpretation of the additional sort names and operators is restricted by initial satisfaction.

We call the condition (i) the persistency and the condition (ii) the *initiality* of an enrichment by definition of a canon.

Roughly speaking, a canon consists of a set of sort names, a set of operators, a set of axioms, and a set of so-called *initial restrictions*. On the textual level the initial restrictions are given by *definition enrichments*. A precise meaning of this notion is given by Definition 4.1.1 and 4.1.2.

It may happen that persistency and initiality together prevent the existence of any initial extension. In this case the initial canon semantically represents the empty class of algebras and is therefore of no practical interest. Since there are no syntactic rules that prevent the writing of initial canons with empty model classes, we have to bear in mind this pathological case, too.

We illustrate the persistency and initiality of an enrichment by definition of a canon by the following incorrect enrichment.

```
FINITE PATHS* is DIRECTED GRAPH with definition
    sorts Path
    oprn nil(Nodes) → Path
        add(w:Path,x:Edges iff p-end(w)=begin(x)) → Path
        p-end(Path) → Path
    axioms w1,w2: Path, x: Edges
        if w2 = add(w1,x) then p-end(w2) = end(x)
end FINITE PATHS*
```

Both initial canons differ only in the set of axioms. If G is any directed graph with $G_{\text{Nodes}} \neq \emptyset$, then there is up to isomorphism exactly one extension $W(G)$ that interprets the additional sort names and operators, and that initially satisfies the additional axiom FINITE PATHS*. But this unique extension $W(G)$ is not a persistent extension of G. At first we obtain exactly one object $\text{nil}^{W(G)}(e)$ in $W(G)_{\text{Path}}$ for each node e in G_{Nodes}. Next we can apply the operation $\text{p-end}^{W(G)}$ to each of these new objects. Initiality means that in this way new nodes are created, i.e. $W(G)_{\text{Nodes}} \supset G_{\text{Nodes}}$. This generation of additional nodes is prevented in the specification FINITE PATHS by the axiom $\text{p-end}(\text{nil}(e)) = e$. Additionally the generation of new nodes implies that the domain of $\text{add}^{W(G)}$ becomes empty since there is no edge x in $W(G)_{\text{Edges}}$ with

$$\text{begin}^{W(G)}(x) = \text{p-end}^{W(G)}(\text{nil}^{W(G)}(e))$$

for any e in $W(G)_{\text{Nodes}}$. Now the operation $\text{nil}^{W(G)}$ can be applied to the created new nodes. This application produces new objects in $W(G)_{\text{Path}}$ which can be used as arguments of the operation $\text{p-end}^{W(G)}$. Applying this operation in this way one produces again new nodes in $W(G)_{\text{Nodes}}$. This process can be applied in an unrestricted number of steps. The resulting set $W(G)_{\text{Nodes}}$ properly includes the set of nodes of the given directed graph. The preceding consideration proves that there is no interpretation of FINITE PATHS* that satisfies both persistency and initiality.

After the discussion of possible conflicts between persistency and initiality we come back to the specification FINITE PATHS. As previously mentioned the creation of additional nodes is prevented by the first axiom. As before, G may denote an arbitrary directed graph. If there are an edge $x_1 \in G_{\text{Edges}}$ and a node $e \in G_{\text{Nodes}}$ with $e = \text{begin}^{G}(x_1)$, then the operation $\text{add}^{W(G)}$ is applicable to the pair $(\text{nil}^{W(G)}(e), x_1)$ and an object of sort 'Path' is produced as the value of the formal expression $\text{add}(\text{nil}(e), x_1)$. If it is evident from context in which algebra we are working, we allow omission of the name of the algebra in the notation of an operation. According to that agreement in the given context $\text{add}(\text{nil}(e), x_1)$ abbreviates $\text{add}^{W(G)}$ $(\text{nil}^{W(G)}(e), x_1)$. If in G_{Edges} there is an edge x_2 not necessarily different from x_1 but satisfying $\text{end}(x_1) = \text{begin}(x_2)$, then the operation add is applicable to $(\text{add}(\text{nil}(e), x_1), x_2)$ and a further path as value of the expression $\text{add}(\text{add}(\text{nil}(e), x_1), x_2)$ will be produced. The applicability of add to $(\text{add}(\text{nil}(e), x_1), x_2)$ is a consequence of the second axiom. By setting $w_2 = \text{add}(\text{nil}(e), x_1)$ and $w_1 = \text{nil}(e)$ we obtain $\text{p-end}(\text{add}(\text{nil}(e), x_1))$ $= \text{end}(x_1)$ and because of $\text{end}(x_1) = \text{begin}(x_2)$ we finally get $\text{p-end}(\text{add}$ $(\text{nil}(e), x_1)) = \text{begin}(x_2)$. The expression $\text{add}(\text{add}(\text{nil}(e), x_1), x_2)$ represents a sequence of two connected edges or in other words a path of length two from $\text{begin}(x_1) = e$ to $\text{end}(x_2)$. The sufficiency of the domain condition of the add-operator guarantees that each finite directed path will be created, whereas the necessity of the domain condition prevents the construction of non-connected sequences of edges.

The initial canon FINITE PATHS contains only those operators that are necessary for the generation of all finite directed paths. For working with this PADT it would be convenient to enrich this PADT, for instance with a function that associates with each path its source, or with a function that calculates the length of each path, or one can define the partial composition of paths that concatenates two paths iff they are connected. The following initial canon is a stepwise enrichment of FINITE PATHS by specifications of the suggested functions.

EXAMPLE 2.1 is FINITE PATHS **with** NAT **with definition**
  oprn length(Path) → Nat
  axioms x: Edges, w_1,w_2: Path, e: Nodes
    length(nil(e)) = zero
    if w_2 = add(w_1,x) **then** length(w_2) = succ(length(w_1))
  with definition
  oprn source(Path) → Nodes
  axioms x: Edges, w_1,w_2: Path, e: Nodes
    source(nil(e)) = e
    if w_2 = add(w_1,x) **then** source(w_2) = source(w_1)
  with definition
  oprn compose(w_1:Path,w_2:Path **iff** p-end(w_1)=source(w_2)) → Path
  axioms w_1,w_2: Path, x: Edges
    compose(w_1,nil(p-end(w_1))) = w_1
    if p-end(w_1) = source(add(w_2,x)) **then**
      compose(w_1,add(w_2,x))=add(compose(w_1,w_2),x) **fi**
end EXAMPLE 2.1

In the first line we join by the key word **with** the initial canon FINITE PATHS and the initial canon NAT, since we need natural numbers as results of the required function length(Path) → Nat. The principle we used for the unique specifications of the functions length, source, and composition is evidently the same as we used in Section 1.1 for functional enrichmets of ADTs. But now the operators do not represent an abstract algorithm over an ADT, now the operators represent *parameterized abstract algorithms* which become abstract algorithms if the parameter structure is actualized by an ADT which specifies a directed graph. The problem of parameter substitution will be discussed in detail at the end of this section in connection with a further structuring principle of initial canons called *modification*.

The possibility of combining of enrichment by definitions and *enrichment by requirements* significantly increases the expressiveness of initial canons. We demonstrate this *mixed enrichment* by the following specification. First we enrich FINITE PATHS by a requirement of an integer evaluation of edges, and in a second enrichment, this time by definition, we use this evaluation for the definition of a function that gives for each path the minimal evaluation over all edges of that path.

EXAMPLE 2.2 **is** FINITE PATHS **with** NAT **with requirement**
oprn capacity(Edges) $\rightarrow$ Nat
with definition
oprn min(Nat,Nat) $\rightarrow$ Nat
Nat $<$ Nat $\rightarrow$ Nat
axioms x,y: Nat
not($x <$ zero)
zero $<$ succ(x)
($x < y$) = (succ(x) $<$ succ(y))
min(x,x) = x)
if $x < y$ **then** min(x,y) = x
with definition
oprn cap(Path) $\rightarrow$ Nat
axioms e: Nodes, x,y: Edges, w_1,w_2: Path
cap(nil(e)) = zero
if e=begin(x) **then** cap(add(nil(e),x))=capacity(x)
if w_2 = add(add(w_1,x),y) **then**
cap(w_2) = min(cap(add(w_1,x)),capacity(y)) **fi**
end EXAMPLE 2.2

This example makes evident the difference between the two kinds of enrichments of initial canons. On a more abstract level the difference is given by the fact that enrichments by definition are characterized by persistency and initiality, whereas for enrichments by requirement only persistency is claimed.

In Section 1.1 the specification of the identity of an ADT was not a trivial task. The strong expressiveness of initial canons makes this problem a trivial one as we can see by the following specification.

EXAMPLE 2.3 **is** FINITE PATHS **with** BOOL **with requirement**
oprn equal(Path,Path) $\rightarrow$ Bool
axioms w,w_1,w_2: Path
equal(w,w) = true
if equal(w_1,w_2) = true **then** $w_1 = w_2$
end EXAMPLE 2.3

This specification of the identity of paths is not based on the inductive definition of finite paths. Even if the operator equal is introduced within an enrichment by requirement, the persistency together with the stated axioms guarantees that this operator can be interpreted only in one way, namely as the intended identity. In this way we can semantically uniquely specify the identity for each sort name of an initial canon.

The equivalent problem to Section 1.1 is now the question whether the identity of paths can be specified by an enrichment by definition of FINITE PATHS. Intuitively it seems to be clear that the answer is no, since there is no possibility to check the identity of edges and of nodes. By means of the results of Section 4.2 we can confirm this intuition. But, if the identity of edges is available, it is not hard to find an inductive definition of the identity of paths. There are several possibilities for such a specification. One of them is the following.

EXAMPLE 2.4 **is** EXAMPLE 2.1 **with requirement**
oprn e-eq(Edges,Edges) → Bool
n-eq(Nodes,Nodes) → Bool
axioms x,x': Edges, e,e': Nodes
e-eq(x,x) = true
if e-eq(x,x') = true **then** $x = x'$
n-eq(e,e) = true
if n-eq(e,e') = true **then** $e = e'$
with definition
oprn equal(Path,Path) → Bool
axioms e,e': Nodes, x,y: Edges, w,w',v,v': Path
equal(nil(e),nil(e')) = n-eq(e,e')
if eq(length(v),length(w)) = false **then** equal(v,w) = false
if v = add(w,x), v' = add(w',y) **then**
equal(v,v') = and(equal(w,w'),e-eq(x,y) **fi**
end EXAMPLE 2.4

In connection with the substitution of parameters it is convenient to assume that for each sort name of an initial canon that is introduced within an enrichment by requirement its equality operator is also introduced. Since those sort names represent parameters, this assumption corresponds to the availability of the identity for each parameter type. One intended consequence of this assumption is that those parameter types can be actualized only by computable ADTs, because the required identity is then also subject to actualization. According to this convention we enrich DIRECTED GRAPH to D-GRAPH:

D-GRAPH **is** DIRECTED GRAPH **with** BOOL **with requirement**
oprn n-eq(Nodes,Nodes) → Bool
e-eq(Edges,Edges) → Bool
axioms x,y: Nodes, a,b: Edges
n-eq(x,x) = true
if n-eq(x,y) = true **then** $x = y$
e-eq(a,a) = true
if e-eq(a,b) = true **then** $a = b$
end D-GRAPH

In the preceding specification BOOL denotes a specification of the ADT of truth values with the usual operations $\wedge$, $\vee$, $\neg$.

Next we study the limits of expressiveness of enrichments by definitions of initial canons. In Section 4.2 we will show that it is justified to consider enrichments by definitions that define functional extensions as specifications of computable functions parameterized by the given initial canon, or as parameterized computable functions for short. Let the given base canon be D-GRAPH and let us consider the parameterized function

$$\text{connected}^G(G_{\text{Nodes}}, G_{\text{Nodes}}) \rightarrow \{\text{true,false}\}$$

where G denotes any model of D-GRAPH. This function may take the value true for an ordered pair (e_1,e_2) of nodes in G if and only if there is a finite directed path over G from e_1 to e_2. Now we can put the question whether this parameterized function is computable over D-GRAPH or not.

In Chapter 4 we prove properties of enrichments by definitions that give an answer to this question. Theorem 4.2.4 in particular gives a necessary condition for computable functions parameterized by an initial canon $\mathbb{B}$. Roughly speaking this theorem states the compatibility of homomorphisms between models of $\mathbb{B}$ with the considered function. With respect to our example this means that for each homomorphism $f\colon A \rightarrow B$ between models A and B of D-GRAPH

$$\text{connected}^A(e_1,e_2) = \text{connected}^B(f_{\text{Nodes}}(e_1), f_{\text{Nodes}}(e_2))$$

for all $e_1, e_2 \in A_{\text{Nodes}}$, if the function 'connected' is computable with respect to D-GRAPH.

Since a homomorphism $f\colon A \rightarrow B$ between models of D-GRAPH has to be compatible with each operator of D-GRAPH (see Section 2.2), the single functions

$$f_{\text{Nodes}}\colon A_{\text{Nodes}} \rightarrow B_{\text{Nodes}}, \quad f_{\text{Edges}}\colon A_{\text{Edges}} \rightarrow B_{\text{Edges}}$$

satisfy the following conditions:

$$f_{\text{Nodes}}(\text{begin}^A(x)) = \text{begin}^B(f_{\text{Edges}}(x))$$

and

$$f_{\text{Nodes}}(\text{end}^A(x)) = \text{end}^B(f_{\text{Edges}}(x)) \quad \text{for each} \quad x \in A_{\text{Edges}},$$

and

$$\text{n-eq}^A(e_1,e_2) = \text{n-eq}^B(f_{\text{Nodes}}(e_1), f_{\text{Nodes}}(e_2)) \quad \text{for all} \quad e_1,e_2 \text{ of } A_{\text{Nodes}},$$

$$\text{e-eq}^A(x_1,x_2) = \text{e-eq}^B(f_{\text{Edges}}(x_1), f_{\text{Edges}}(x_2)) \quad \text{for all edges} \quad x_1,x_2 \in A_{\text{Edges}}.$$

These conditions imply that for each homomorphism $f\colon A \rightarrow B$ between models of D-GRAPH the functions $f_{\text{Nodes}}\colon A_{\text{Nodes}} \rightarrow B_{\text{Nodes}}$ and $f_{\text{Edges}}\colon A_{\text{Edges}} \rightarrow B_{\text{Edges}}$ are one-to-one.

For the given example it is now easy to find models A and B of D-GRAPH and to construct a homomorphism $f: A \to B$ that is not compatible with the parameterized function

$$\text{connected(Nodes,Nodes)} \to \text{Bool}.$$

We can take models A and B of D-GRAPH with $A_{\text{Nodes}} = B_{\text{Nodes}} = M$ an arbitrary set, with $A_{\text{Edges}} = \emptyset$ the empty set, with B_{Edges} any non-empty set, and with any functions begin^B, end^B. Evidently we set $A_{\text{Bool}} = B_{\text{Bool}} = \{\text{true, false}\}$.

With this assumption we obtain a homomorphism $f: A \to B$ by setting $f_{\text{Nodes}} = \text{Id}_M$ and taking $f_{\text{Edges}}: \emptyset \to B_{\text{Edges}}$ the unique inclusion of the empty set, and by $f_{\text{Bool}} = \text{Id}_{\{\text{true,false}\}}$.

Because of $A_{\text{Edges}} = \emptyset$ any two nodes of A are inconnected. Now we can choose B_{Edges}, begin^B, and end^B in such a way that there are at least two $e_1, e_2 \in B_{\text{Nodes}} = A_{\text{Nodes}}$ that are connected in B by a finite directed path from e_1 to e_2. For such a choice we obtain

$$\text{connected}^A(e_1, e_2) = \text{false},$$

$$\text{connected}^B(f_{\text{Nodes}}(e_1), f_{\text{Nodes}}(e_2)) = \text{connected}^B(e_1, e_2) = \text{true},$$

so that this homomorphism is not compatible with the considered function. This shows that the class of models of D-GRAPH is much too rich to allow a definition of the intended parameterized function, or in other words there are not enough fundamental operations for a definition of the connected-operator.

One can summarize the preceding considerations in the phrase that the *computational capacity* of the initial canon D-GRAPH is too weak for a specification of the intended function by means of enrichments by definitions. Perhaps this result is not too surprising because we permit any directed graph as parameter. What about the restriction to directed graphs with finite sets of nodes and edges? Would this restriction sufficiently increase the computational capacity? Before we try to answer this question we have to answer another one. Is it possible to represent the class of all finite directed graphs as the class of all models of a suitable initial canon? Only if the answer to that question is yes does the previous question become meaningful. By the following specification we demonstrate that the class of all finite directed graphs is representable as a class of models of an initial canon, although that class is beyond the range of first-order logic.

FINITE DIRECTED GRAPH **is** D-GRAPH **with definition**
  sorts Set
  oprn empty → Set
    e-add(Set,Edges) → Set
    n-add(Set,Nodes) → Set

axioms a, a_1, a_2: Nodes, x, x_1, x_2: Edges, s: Set
e-add(e-add(s,x),x) = e-add(s,x)
n-add(n-add(s,a),a) = n-add(s,a)
e-add(e-add(s,x_1),x_2) = e-add(e-add(s,x_2),x_1)
n-add(n-add(s,a_1),a_2) = n-add(n-add(s,a_2),a_1)
e-add(n-add(s,a),x) = n-add(e-add(s,x),a)
with requirement
oprn edges $\rightarrow$ Set
nodes $\rightarrow$ Set
axioms a: Nodes, x: Edges
e-add(edges,x) = edges
n-add(nodes,a) = nodes
end FINITE DIRECTED GRAPH

For each model A of FINITE DIRECTED GRAPH with $A_{\text{Edges}} \cap A_{\text{Nodes}} = \emptyset$ the set A_{Set} consists of all finite subsets of $A_{\text{Edges}} \cup A_{\text{Nodes}}$, $\text{empty}^A \in A_{\text{Set}}$ is the empty set, and the operations e-add^A, n-add^A are given by

$$\text{e-add}^A(s,x) = s \cup \{x\}, \quad \text{and} \quad \text{n-add}^A(s,a) = s \cup \{a\}$$

for each finite subset $s \in A_{\text{Set}}$ and each $x \in A_{\text{Edges}}$, $a \in A_{\text{Nodes}}$. The assumed disjointness of A_{Edges} and A_{Nodes} does not restrict generality. In this case A_{Set} becomes the set of all finite subsets of the disjoint union of A_{Edges} and A_{Nodes}. Due to the initiality and persistency of the enrichment by definition the interpretation of Set, empty, e-add, and n-add is unique up to isomorphism. However, the restriction to finite directed graphs does not result from the enrichment by definition but from the enrichment by requirement. This enrichment requires the distinction of two finite subsets

$$\text{edges}^A \in A_{\text{Set}}, \qquad \text{nodes}^A \in A_{\text{Set}}$$

such that each node $a \in A_{\text{Nodes}}$ is an element of nodes^A, and each edge $x \in A_{\text{Edges}}$ is an element of edges^A. A distinction of those elements in A_{Set} is practicable only for a finite directed graph A.

After that explanation of the specification FINITE DIRECTED GRAPH we come back to the question whether the restriction to finite directed graphs sufficiently increases the computational capacity. However, the same construction of a homomorphism $f: A \rightarrow B$ between directed graphs that has been used before can now also be used for an answer in the negative, since each homomorphism $f: A \rightarrow B$ between finite directed graphs can be uniquely extended to a homomorphism between models of FINITE DIRECTED GRAPH. This negative answer indicates that a sufficient increase of the computational capacity requires additional operations.

There are several ways of sufficiently increasing the computational capacity of D-GRAPH. But we will not attend to that matter; instead we will show that the intended function can be *uniquely specified* by a mixed enrichment.

```
C-D-GRAPH is D-GRAPH with definition
    sorts Path
    oprn nil(Nodes) → Path
         add(w:Path,x:Edges iff p-end(w)=begin(x)) → Path
         p-end(Path) → Nodes
         p-begin(Path) → Nodes
    axioms e: Nodes, x: Edges, w1,w2: Path
         p-end(nil(e)) = e
         p-begin(nil(e)) = e
         if w2 = add(w1,x) then p-end(w2) = end(x)
         if w2 = add(w1,x) then p-begin(w2) = p-begin(w1)
    with requirement
    oprn connected(Nodes,Nodes) → Bool
         path(x:Nodes,y:Nodes iff connected(x,y)=true) → Path
    axioms x,y: Nodes, w: Path
         if x=p-begin(w), y=p-end(w) then connected(x,y) = true
         if connected(x,y) = true then x = p-begin(path(x,y))
         if connected(x,y) = true then y = p-end(path(x,y))
end C-D-GRAPH
```

Both C-D-GRAPH and FINITE DIRECTED GRAPH clearly demonstrate the profound difference in the expressiveness of enrichments by definitions and of mixed enrichments.

The previous discussion was concerned with the limits of expressiveness of specifications. In the following we demonstrate the expressiveness of enrichments by definition if partial operations are available in contrast to specifications with total operations only. The following specification of the semantics of a *loop-statement*, for instance of the form

while p **do** g **else** f

will be used to prove the adequacy of structural induction on partial algebras. We assume that $p\colon M \hookrightarrow \{\text{true,false}\}$ is a partial predicate on a parameter set M, and that $f,g\colon M \hookrightarrow M$ are partial transformations.

```
WHILE is BOOL with requirement
    sorts M
    oprn p(M) ⇀ Bool
         f(M) ⇀ M
         g(M) ⇀ M
    with definition
    oprn while(M) ⇀ M
```

axioms $x: M$
    if $p(x)=\text{false}, f(x)=f(x)$ **then** $\text{while}(x) = f(x)$
    if $p(x)=\text{true}, \text{while}(g(x))=\text{while}(g(x))$ **then**
        $\text{while}(x) = \text{while}(g(x))$ **fi**
end WHILE

For a correct understanding of this specification one should have in mind that applicability requirements can be expressed by axioms. In the specification WHILE the domain of the operator 'while' is defined by means of domain requirements resulting only from both axioms. The first axiom states that an element x is in the domain of 'while' whenever x is in the domain of p and in the domain of f and if additionally x does not satisfy p. In that case the first axiom additionally fixes the value $\text{while}(x) = f(x)$. The first axiom represents the basis of the inductive definition. The induction step is given by the second axiom. If we assume that x is in the domain of p, that x satisfies p and that x is in the domain of the composed function $g \circ \text{while}: M \multimap M$, i.e. that x is in the domain of g and $g(x)$ is in the domain of 'while', then we are able to conclude from the second axiom that x is in the domain of 'while', too. We can determine the value of while(x) if we assume that $g(x)$ is in the domain of 'while' because of the first axiom. In that case we would obtain $\text{while}(x) = \text{while}\big(g(x)\big) = f\big(g(x)\big)$. If we continue in this way we obtain that x is in the domain of 'while' if and only if there is a natural number n such that $p(x), p\big(g(x)\big), \ldots, p\big(g^n(x)\big)$, $f\big(g^n(x)\big)$ are all defined, and that $p(x) = p\big(g(x)\big) = \ldots = p\big(g^{n-1}(x)\big) = \text{true}$, and $p\big(g^n(x)\big) = \text{false}$. Then $\text{while}(x) = f\big(g^n(x)\big)$ holds.

The initiality of the enrichment by definition guarantees that the domain of the operator 'while' does not contain additional elements. We hope to improve the understanding by the following discussion of a slightly modified but incorrect specification of the semantics of the while-statement. This specification WHILE* may result from WHILE by a substitution of the equation $\text{while}\big(g(x)\big) = \text{while}\big(g(x)\big)$ in the premise of the second axiom by the equation $g(x) = g(x)$.

We will now construct the domain of the partial operator $\text{while}(M) \multimap M$ but now defined by WHILE*. Since the first axiom is not modified, again an element x is in the domain of the operator 'while' if $p(x)$ is defined and gives the value false, and if additionally $f(x)$ is defined. If x is an element for which $p(x)$ and $g(x)$ are defined and $p(x) = \text{true}$, then the modified version of the second axiom implies that the partial operator 'while' is defined for that x and for $g(x)$. Let M be for instance the set of natural numbers, $p(x) = \text{true}$ iff $x \neq 0$, $g(x) = x + 1$ for each $x \in \mathbb{N}$ and $f: \mathbb{N} \multimap \mathbb{N}$ any partial function defined on zero. For this actual parameter the specification WHILE defines the partial function $w: \mathbb{N} \multimap \mathbb{N}$ with $\text{dom}(w) = \{0\}$ and $w(0) = f(0)$. If we start with $x = 1$, then the first axiom of WHILE is not immediately applicable because of $p(x) = \text{true}$. According to the second axiom of WHILE, while(1) is defined if while(1+1) is defined. If we want to prove that while(2) is defined, we have to prove that while(2+1) is defined and this goes on infinitely. This shows that for no natural number x with $x \neq 0$ is there a proof of the applicability of 'while' to x on the basis of the given axioms of WHILE. But if we are looking for a proof of the applicability of 'while' to x with $x \neq 0$ on the basis of the axioms of WHILE*, the situation is completely different.

Since $g(1)$ is defined we now obtain that both while(1) and while$(g(1))$ are defined and that both values are equal, i.e. while(1) = while(2). In the same way we obtain for any $x \neq 0$ that while(x) and while$(g(x))$ = while(x+1) are defined and equal. Evidently there are many total functions while: $\mathbb{N} \to \mathbb{N}$ with while(0) = f(0) and while(1) = while(2) = ... = while(x) = ... Since there is no axiom that implies the equality of while(1) with any natural number, the initiality of the enrichment by definition in WHILE* causes the creation of a new object of sort M. The creation of a new object contradicts the persistency so that not each actual parameter of WHILE* can be extended to a model of WHILE*.

We hope that the similarity of the axioms in WHILE with the ordinary explanation of the semantics of a while-statement has been observed. We interpret this similarity as a justification of the introduced calculus of functional enrichments of PADTs.

Before we will deal with the *actualization of formal parameters* of an initial canon we specify a further example for which the class of actual parameters is of greater interest than in the previous specifications.

The object of the specification is some kind of an inhomogeneous storage. For this inhomogeneous storage besides the sets of addresses and values a set of types and some additional functions may be given that associate with each address and with each value an object of sort 'Type'. In addition the quality of addresses may be given. This system of sets and functions is considered as a parameter of the inhomogeneous storage. On the basis of that parameter we define the set of all states of the storage and we define store-, read-, and remove-operations. For these operations we assume that a value x can be stored by an address a only if a and x are both of the same type. Evidently, one can only read a value by an address, if under that address some value has previously been stored. These requirements imply that both operations become partial ones. The remove-operation will be specified as a total operation that will not change the state of the storage if we try to remove an information by an address under which before no value has been stored.

STORAGE **is** BOOL **with requirement**
  sorts Address, Value, Type
  oprn a-type(Address) → Type
    v-type (Value) → Type
    equal(Address,Address) → Bool

axioms a, a_1, a_2: Address
    equal(a,a) = true
    if equal(a_1,a_2) = true **then** $a_1 = a_2$
with definition
sorts Storage
oprn initiate → Storage
    written(Address,Storage) → Bool
    store(Storage,a:Address,x:Value
        iff a-type(a) = v-type(x)) → Storage
    read(a:Address,s:Storage
        iff written(a,s) = true) → Value
    remove(Address,Storage) → Storage
axioms a, a_1, a_2: Address, s: Storage, x,y: Value
    written(a,initiate) = false
    if written (a_1,s) = false, a-type(a_2) = v-type(x) **then**
        written(a_1,store(s,a_2,x)) = equal(a_1,a_2) **fi**
    if a-type(a)=v-type(x), a-type(a)=v-type(y) **then**
        store(store(s,a,x),a,y) = store(s,a,y) **fi**
    if a-type(a) = v-type(x), a-type(a_2) = v-type(y),
        equal(a ,a_2) = false **then**
        store(store(s,a ,x),a_2,y) = store(store(s,a_2,y),a ,x) **fi**
    if a-type(a) = v-type(x) **then** remove(a,store(s,a,x))
        = remove(a,s) **fi**
    if written(a,s) = false **then** remove(a,s) = s
    if a-type(a)=v-type(x) **then** read(store(s,a,x)) = x
end STORAGE

In this specification explicit domain conditions within the declaration of the partial operators are used as distinct from the initial canon WHILE. Therefore we take care that no additional domain conditions are caused by the axioms.

By means of the actualization of the parameter part one can derive different kinds of storages. It is important that actual parameters themselves may be PADTs. This possibility enables *stepwise refinements* of PADTs.

In the following the abstract storage will be actualized by setting the disjoint union of the sets of natural numbers, truth values, and strings over an arbitrary alphabet for the parameter sort 'Value'. The parameter sort 'Address' will be actualized by a set of strings over a possibly different alphabet, and we will finally set the two-element set {static,dynamic} for the parameter sort 'Type'. Therefore the attribute functions 'a-type' and 'v-type' associate with each address and each value one of each of these attributes. This actualization uses another parameterized abstract data type as actual parameter so that by further actualizations both the set of characters of the strings that form the addresses and the set of characters that form a part of the values may be actualized. To achieve the sketched actualization we specify first the actual parameter.

PARAMETER1 is NAT **with requirement**
sorts Character
oprn c-eq(Character,Character) → Bool
axioms x,y: Character
c-eq(x,x) = true
if c-eq(x,y) = true **then** $x = y$
with definition
sorts String
oprn nil → String
append(String,Character) → String
join(String,String) → String
s-eq(String,String) → Bool
axioms x,y: Character, u,v: String
join(v,nil) = v
join(u,append(v,x)) = append(join(u,v),x)
s-eq(u,u) = true
s-eq(u,v) = s-eq(v,u)
s-eq(u,join(u,append(v,x))) = false
s-eq(append(u,x),append(v,y)) = and(s-eq(u,v),c-eq(x,y))
with definition
sorts Union, Kind
oprn static, dynamic → Kind
b-inj(Bool) → Union
n-inj(Nat) → Union
s-inj(String) → Union
u-attr(Union) → Kind
axioms x: Bool, y: Nat, v: String
u-attr(b-inj(x)) = static
u-attr(n-inj(y)) = static
u-attr(s-inj(v)) = dynamic
end PARAMETER 1

For an actualization of the parameter sort 'Character' in the initial canon PARAMETER 1 we define a special set of characters by the initial canon PARAMETER 2.

PARAMETER2 is BOOL **with definition**
sorts A
oprn a,b,c → A
a-eq(A,A) → **Bool**

axioms $x: A$

a-eq(x,x) = true

a-eq(a,b) = a-eq(a,c) = a-eq(b,a) = a-eq(b,c) = a-eq(c,a)

= a-eq(c,b) = false

end PARAMETER 2

After the specification of a PADT and of an ADT which can be used as actual parameter afterwards, we introduce the last operation for the construction of initial canons. This construction is called a *modification*, and is based on the parameterization concept suggested by ZILLES (1981).

$\mathfrak{C}$ **is** $\mathfrak{C}_1$ **modified** [**by** $\mathfrak{C}_2$] **according**

S_1 **for** S_1^*

....

s_n **for** S_n^* (+)

σ_1 **for** σ_n^*

....

σ_m **for** σ_m^*

end $\mathfrak{C}$

such that:

(i) $S_1^*, \ldots, S_n^*$ are sort names of the initial canon $\mathfrak{C}_1$;

(ii) $\sigma_1^*, \ldots, \sigma_m^*$ are operators of the initial canon $\mathfrak{C}_1$;

(iii) S_i and σ_j for $i \in \{1, \ldots, n\}$, $j \in \{1, \ldots, m\}$ are either sort names and operators of $\mathfrak{C}_2$ or they are new names of sorts and operators different from all sort names and operators of $\mathfrak{C}_1$ and $\mathfrak{C}_2$ respectively;

(iv) The *modification list*, i.e. the part between **according** and **end**, is compatible with the operator declarations in $\mathfrak{C}_1$ and $\mathfrak{C}_2$, i.e. if $\sigma_i^*(S_{i_1}^*, \ldots, S_{i_k}^*) \to S_{i_0}^*$ is an operator declaration in $\mathfrak{C}_2$ then the associated operator in $\mathfrak{C}_1$ is declared as follows: $\sigma_i(S_{i_1}, \ldots, S_{i_k}) \to S_{i_0}$. The compatibility of the modification list with the operator declarations in $\mathfrak{C}_1$ and $\mathfrak{C}_2$ includes the requirement that for every operator $\sigma_i^*(S_{i_1}^*, \ldots, S_{i_k}^*) \to S_{i_0}^*$ contained in the right-hand side of the modification list each sort name $S_{i_j}^*$, $j = 0, 1, 2, \ldots, k$, is either a common sort name of $\mathfrak{C}_1$ and $\mathfrak{C}_2$ or it is contained in the right-hand side of the modification list.

(v) The modification list defines a canon morphism $\mu: \mathfrak{C}_0 \to \mathfrak{C}_2$ from the subcanon $\mathfrak{C}_0$ induced by the subtheory T_0 of the theory T_1 of $\mathfrak{C}_1 = (T_1, \Delta_1)$, where T_0 is given by $\{S_1^*, \ldots, S_n^*\} = S(T_0)$, $\{\sigma_1^*, \ldots, \sigma_m^*\} = \Omega(T_0)$, and where $\mathfrak{A}(T_0)$ is the subset of the set $\mathfrak{A}(T_1)$ of axioms of the canon $\mathfrak{C}_1$ consisting of all axioms containing only operators of $\{\sigma_1^*, \ldots, \sigma_m^*\}$.

The initial canon $\mathfrak{C}_1$ is called the *modified canon* of the modification (+), and the initial canon $\mathfrak{C}_2$ is called the *modifying canon*. As indicated by rectangular brackets the part 'by $\mathfrak{C}_2$' of the modification (+) is optional.

If it is omitted the left-hand side of the modification list can contain only new names of sorts and new operators, since in this case the modification is reduced to a simple renaming of some sort names and some operators of $\mathbb{C}_1$. Formally the initial canon $\mathbb{C}$ resulting from a modification (+) of $\mathbb{C}_1$ by $\mathbb{C}_2$ is constructed by the following *pushout*:

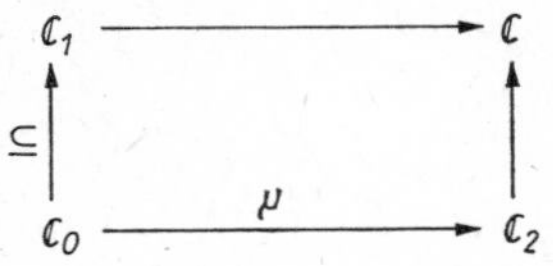

where $\mu: \mathbb{C}_0 \to \mathbb{C}_2$ denotes the canon morphism given by the modification list according to condition (v). For a precise definition of a subcanon induced by a subtheory and of the pushout construction of canon morphisms see Definition 4.3.1 and Theorem 4.3.3.

Roughly speaking, the initial canon $\mathbb{C}$ resulting from a modification of $\mathbb{C}_1$ by $\mathbb{C}_2$ according to a given modification list is constructed in the following way: the subcanon $\mathbb{C}_0$ of $\mathbb{C}_1$ induced by $S_1^*, \ldots, S_n^*$ and $\sigma_1^*, \ldots, \sigma_m^*$ will be substituted by the image $\mu(\mathbb{C}_0)$ and the resulting initial canon will be joined with the initial canon $\mathbb{C}_2$. Using a modification one should be careful to avoid name conflicts.

Since every enrichment by definition of $\mathbb{C}_0$ is transformed by the canon morphism $\mu: \mathbb{C}_0 \to \mathbb{C}_2$ into a corresponding enrichment by definition in $\mathbb{C}_2 \subseteq \mathbb{C}$, it is impossible to change an enrichment by definition into an enrichment by requirement by means of a modification. The reverse is the case: the aim of modifications is to transform enrichments by requirements in $\mathbb{C}_1$ into enrichments by definition in the resulting initial canon $\mathbb{C}$. (For the concepts of theory, canon, and canon morphism see (2.5.5), (4.1.2), and (4.3.1), respectively.) This implies that each modification of an initial canon that does not contain any enrichment by requirement can be only a renaming of a subcanon of $\mathbb{C}_1$ followed by a union with $\mathbb{C}_2$. However, this situation should be made more explicit by a construction like

$\mathbb{C}$ **is** $\mathbb{C}_1$ **modified according** ⟨modification list⟩ **with** $\mathbb{C}_2$.

The modification

PARAMETER 3 **is** PARAMETER1 **modified by** PARAMETER 2
according
    A **for** Character
    a-eq **for** e-eq
end PARAMETER 3

transforms the PADT given by the specification PARAMETER1 into the specification PARAMETER 3 of an ADT, since the enrichment by requirement in PARAMETER1 is transformed into an enrichment by definition. We will demonstrate the effect of a modification by this example. The subcanon $\mathbb{C}_0$ of PARAMETER1, generated by Characters and e-eq is given by:

> $\mathbb{C}_0$ **is** BOOL **with requirement**
>   **sorts** Character
>   **oprn** c-eq(Character,Character) → Bool
>   **axioms** x,y: Character
>   c-eq(x,x) = true
>   **if** c-eq(x,y)=true **then** $x=y$
> **end** $\mathbb{C}_0$

Even if the modification list does not contain the sort name 'Bool' this sort name is contained in $\mathbb{C}_0$ because of the occurrence of the operator 'c-eq' in the modification list. The image $\mu(\mathbb{C}_0)$ in the specification PARAMETER 2 is the initial canon PARAMETER 2 itself. This implies that $\mathbb{C}_1^*$ becomes $\mathbb{C}$, so that we obtain the following specification as the result of the modification:

> PARAMETER 3 **is** NAT **with definition**
>   **sorts** A
>   **oprn** $a,b,c \to A$
>   a-eq(A,A) → Bool
>   **axioms** x: A
>   a-eq(x,x) = true
>   a-eq(a,b) = a-eq(a,c) = a-eq(b,a) = a-eq(b,c) = a-eq(c,a) =
>   = a-eq(c,b) = false
>   **with definition**
>   **sorts** String
>   **oprn** nil → String
>   append(String,A) → String
>   s-eq(String,String) → Bool
>   **axioms** x,y: A, u,v: String
>   ....
>   the same continuation as in PARAMETER1
>   ....
> **end** PARAMETER 3

We obtain a prototype of the ADT specified by PARAMETER 3 if we take for 'Nat' the set of natural numbers, for 'Bool' the set of truth values, for 'A' the three-element alphabet $\{a,b,c\}$, for 'String' the set $\{a,b,c\}^*$ of all finite sequences of elements of $\{a,b,c\}$, for 'Union' the set $\mathbb{N} \cup$ {true,false} $\cup \{a,b,c\}^*$, and for 'Kind' the two-element set {static,dynamic}. The semantic interpretation of the operators can easily be deduced from the names of the operators.

The following modification of STORAGE is a first step in the direction to the intended actualization.

STORAGE1 **is** STORAGE **modified by** PARAMETER1 **according**
    Union **for** Value
    Kind **for** Type
    u-attr **for** v-type
end STORAGE1

An expanded version of the initial canon STORAGE1 is given by the following specification.

STORAGE1 **is** PARAMETER1 **with requirement**
    sorts Address
    oprn a-type(Address) → Kind
        equal(Address,Address) → Bool
    axioms a,a_1,a_2: Address
        equal(a,a) = true
        if equal(a_1,a_2) = true **then** $a_1 = a_2$
    with definition
    ....
        the same continuation as in STORAGE
    ....
end STORAGE1

In this specification the sort names 'Character' and 'Address' represent parameters. Both sort names may be interpreted by arbitrary sets. The intended actualization of STORAGE will be obtained by a modification of STORAGE1 using the following modifying canon PARAMETER 4.

PARAMETER4 **is requirement**
    sorts Sign
    with definition
    sorts Sequence
    oprn nil → Sequence
        add(Sequence,Sign) → Sequence
end PARAMETER 4

The described actualization of STORAGE is then given by

STORAGE 2 **is** STORAGE 1 **modified by** PARAMETER 4 **according**
    Sequence **for** Address
end STORAGE 2

The following modification of STORAGE yields a more restricted actualization than the previous one by restricting addresses to natural numbers.

STORAGE 3 is STORAGE modified by PARAMETER 1 according
Union **for** Value
Kind **for** Type
u-attr **for** v-type
Nat **for** Address
eq **for** equal
end STORAGE 3

However, STORAGE 3 does not specify an ADT since the sort name 'Character' and the operator 'a-type(Nat) → Kind' represent parameters.

With these examples we conclude Section 1.2, which is intended to be an informal representation of the different means of writing specifications that may contain partial operations.

It is not the aim of Section 1.2 to give a general discussion of parameterizations of specifications. The aim of this section is the demonstration of the expressive power of structural induction on equationally partial algebras making possible the semantics definition of recursive functionals like the 'while'-construction independent from fixed point theory, introduced by D. Scott, in a completely abstract algebraic way but in a very intuitive style.

With regard to the specification of recursive functionals, or higher-order functions called in Standard ML, the concept of *enrichments by requirement* serves not only for specifications with parameters. Just as polymorphic types and higher-order functions are interesting in themselves, initial specifications with requirement enrichments are useful in themselves.

Concerning the different constructions of parameterized specifications we believe that there is no single one that is to be preferred. It seems to be a very suitable approach to define a *kernel language* that allows a user to define his own specification building operations, or to define his own high-level specification language. This approach was suggested by Sannella and Wirsing (1983). An interesting application of this idea is given by Sannella and Tarlecki (1985).

1.3. The definition of the semantics of a functional language over arbitrary PADTs as a complex example

The main point of this section is the specification of a complex example. But the search for a suitable example is not a simple task. Examples for which there are mathematical descriptions do not require formal specifications. Examples for which there are not adequate mathematical descriptions are hard to explain independently of a formal specification. Therefore we will use an example that can roughly be represented in an informal way, but a detailed and unique representation will require formal specification. The definition of the semantics of a *functional language* over an arbitrary PADT, given by an initial canon $\mathbb{B}$, will serve as such an example.

Since the specification is parameterized by a many-sorted signature, i.e. by a family of predefined types and a family of predefined functions, only the choice of an actual signature and of an actucal algebra of that signature would complete the semantics to the semantics of an actual functional language. Actually we specify the semantics of a whole family of functional languages. In terms of the programming language ML the sort names of the parameter signature represent polymorphic types and the operators of the parameter signature represent polymorphic functions. The intended functional language does not contain facilities for the definition of new data types. Actually we will define the semantics so that we restrict the choice of the algebra to the model class of any chosen initial canon $\mathbb{B}$. Due to this restriction we will speak of $\mathbb{B}$-types, given by the sort names of $\mathbb{B}$, and of $\mathbb{B}$-functions, given by operators of $\mathbb{B}$ or of suitable definition enrichments $\mathbb{B}^*$ of $\mathbb{B}$. Since $\mathbb{B}$ is an arbitrary initial canon, the effect of the definition of a new data type can be achieved by taking a properly enriched canon $\mathbb{B}^*$ as new base canon.

The functional language is basically given by the following five higher-order $\mathbb{B}$-function, $\mathbb{B}$-functionals for short:

composition
looping
branching
source tupling
projection

By means of these $\mathbb{B}$-functionals one can construct new $\mathbb{B}$-functions from the predefined $\mathbb{B}$-functions.

Generally we assume that a $\mathbb{B}$-function f takes an n-tuple ($n \geq 0$) of arguments and if defined for that n-tuple it returns an m-tuple ($m \geq 0$) of results, whereas the predefined $\mathbb{B}$-functions are assumed to return only one result. Notice that we deal with partial $\mathbb{B}$-functions. A declaration of a $\mathbb{B}$-function would look like

$$f\colon S_1 * S_2 * \ldots * S_n \to R_1 * R_2 * \ldots * R_m$$

where $S_1, S_2, \ldots, S_n, R_1, \ldots, R_m$ are $\mathbb{B}$-types, i.e. sort names of $\mathbb{B}$.

The composition of $\mathbb{B}$-functions is a partial $\mathbb{B}$-functional, since the $\mathbb{B}$-functions $f\colon u \circ\!\!\to v$, $g\colon v' \circ\!\!\to w$ (where u,v,v',w here and in the following denote finite sequences of $\mathbb{B}$-types) can be composed if and only if $v = v'$. The result is a $\mathbb{B}$-function

$$f; g\colon u \circ\!\!\to w.$$

The looping requires three arguments $p\colon u \circ\!\!\to \text{Bool}$, $f\colon u \circ\!\!\to u$, $g\colon u \circ\!\!\to u$ and corresponds to the well-known construction

$$\text{while } p \text{ do } g \text{ else } f\colon u \circ\!\!\to u.$$

Using the predefined $\mathbb{B}$-type 'Bool' means that it is assumed that the initial cannon $\mathbb{B}$ contains a specification of the truth values. The arguments of the looping have to satisfy the following conditions: The input sequence of all three arguments are equal and the output sequences of f and g are equal to the common input sequence. The output sequence of p is the sequence of length one consisting of the predefined $\mathbb{B}$-type 'Bool'.

The branching construction corresponds to another well-known construction of programming languages, namely to

$$\text{if } p \text{ then } g \text{ else } f: u \hookrightarrow v$$

where $p: u \hookrightarrow \text{Bool}$, $g: u \hookrightarrow v$, $f: u \hookrightarrow v$ are the arguments. For this $\mathbb{B}$-functional it is required that p, f, g have a common input sequence, that f and g have a common output sequence which may differ from the input sequence, and that p has the output sequence 'Bool'.

The source tupling is applicable to a finite sequence of $\mathbb{B}$-functions

$$f_1: u \hookrightarrow v_1, \quad f_2: u \hookrightarrow v_2, \ldots, \quad f_n: u \hookrightarrow v_n, \quad n \geq 0,$$

with a common input sequence. The result is a $\mathbb{B}$-function

$$(f_1, \ldots, f_n): u \hookrightarrow v_1 * v_2 * \ldots * v_n$$

where the output sequence $v_1 * v_2 * \ldots * v_n$ is the concatenation of the output sequences of the arguments. Semantically the $\mathbb{B}$-function

$$(f_1, \ldots, f_n): u \hookrightarrow v_1 * v_2 * \ldots * v_n$$

represents the parallel execution of $f_1, f_2, \ldots, f_n$. The domain of $(f_1, \ldots, f_n)$ is the intersection of the domains of the single arguments. The source tupling of the empty sequence of $\mathbb{B}$-functions with common input sequence u will be denoted by

$$(\,): u \rightarrow \lambda$$

where λ denotes the empty sequence of $\mathbb{B}$-types. Semantically $(\,): u \rightarrow \lambda$ corresponds to the forgetting of each argument.

Finally, for each sequence of sort names of $\mathbb{B}$, $u = S_1 * S_2 * \ldots * S_n$, and every natural number $i \leq n$, there is a $\mathbb{B}$-function

$$p_i: u \rightarrow S_i$$

picking out the i-th element. p_i will also be called the *i-th projection.*

The semantics of the $\mathbb{B}$-functionals, 'composition', 'source tupling' and 'projection', could be defined without any enrichment by definition of $\mathbb{B}$ simply by working with finite sequences of terms over the signature of $\mathbb{B}$. However, the semantics definition of 'looping' and 'branching' requires enrichments by defintions of $\mathbb{B}$ defining new $\mathbb{B}$-functions which are not expressible as terms or sequences of terms over the signature of the predefined $\mathbb{B}$-functions.

For the precise notion of a term over a many-sorted signature see Section 2.2 If $w = S_1 * S_2 * \ldots * S_n$ is any sequence of sort names of the signature Σ of $\mathbb{B}$, then $T(\Sigma, w)$ denotes the set of all terms (or formal expressions) built of operators of Σ and of exactly one variable for each $\mathbb{B}$-type $S_1, S_2, \ldots, S_n$. If one $\mathbb{B}$-type S appears several times in the sequence w, then we have as many variables of type S as appearances of S in w. With each term $t \in T(\Sigma, w)$ there is uniquely associated a $\mathbb{B}$-type R as result type, namely the result of the operator that has been applied last in that term t. This will be denoted by $t: w \multimap R$.

For any sequence $t_1: w \multimap R_1, \ldots, t_m: w \multimap R_m$ we will write for short

$$\boldsymbol{t}: w \multimap R_1 * R_2 * \ldots * R_m \quad \text{or} \quad \boldsymbol{t}: w \multimap v$$

where $v = R_1 * R_2 * \ldots * R_m$ denotes the output of the term sequence. The i-th element of that sequence will be denoted by

$$\boldsymbol{t}(i): w \multimap R_i \quad \text{for} \quad 1 \leq i \leq m.$$

In Section 2.3 it is shown that any partial algebra A of signature Σ uniquely defines a partial mapping

$$\boldsymbol{t}^A: A_{S_1} \times A_{S_2} \times \cdots \times A_{S_n} \multimap A_{R_1} \times A_{R_2} \times \cdots \times A_{R_m}.$$

Therefore we can formally represent a $\mathbb{B}$-function by an ordered pair

$$(\hat{\mathbb{B}}, \boldsymbol{t}: w \multimap v)$$

where $\hat{\mathbb{B}}$ is an enrichment by definition of $\mathbb{B}$ and $\boldsymbol{t}: w \multimap v$ is a distinguished sequence of terms over the signature of $\hat{\mathbb{B}}$.

Since we need sequences of term sequences, too, and we have to distinguish both of these levels of sequences, we use the notation

$$\boldsymbol{t}^{\#}: w \multimap u_1 \# u_2 \# \ldots \# u_r, \qquad r \geq 0,$$

to denote the following sequence of term sequences, which we will call a $\mathbb{B}$-funtion sequence for short:

$$\left(\boldsymbol{t}^{\#}(1): w \multimap u_1, \quad \boldsymbol{t}^{\#}(2): w \multimap u_2, \ldots, \boldsymbol{t}^{\#}(r): w \multimap u_r\right).$$

Indexing may be used on both levels so that

$$\left(\boldsymbol{t}^{\#}(i)\right)(j): w \multimap u_i(j)$$

denotes the j-th component of the term sequence $\boldsymbol{t}^{\#}(i): w \multimap u_i$.

The semantics of the functional language will now be defined in the following way: For any fixed initial canon $\mathbb{B}$ we define an equationally partial algebra $A(\mathbb{B})$, but we will denote this algebra only by A. It satisfies all axioms of FUNCTIONS, but it is not necessary that A initially satisfies these axioms. Such an algebra A defines a homomorphism

$$\Phi: I \to A$$

for every algebra I that initially satisfies the axioms of FUNCTIONS and that interprets the sort names and operators of SIGNATURE by the same sets and functions as A. FUNCTIONS and SIGNATURE refer to the specifications given in the following.

We demonstrate the expressiveness of the sketched functional language over PADTs by some examples. For these examples we set $\mathbb{B} = \text{NAT}$. If $r: \mathbb{N} \times \mathbb{N} \to \mathbb{N}$ denotes the binary total function with

$$x = y \cdot q + r(x, y), \quad 0 \leq r(x, y) \leq y, \quad \text{and} \quad y \neq r(x, y),$$

and if $p_1: \mathbb{N} \times \mathbb{N} \to \mathbb{N}$, $p_2: \mathbb{N} \times \mathbb{N} \to \mathbb{N}$ denote the projetions to the first and second argument respectively, then the Euclidian algorithm can be expressed by

$$\text{while } p_2 = 0 \text{ do } (p_2, r) \text{ else } (p_1, p_2); p_1. \tag{*}$$

Here $p_2 = 0$ stands for

$$p_2; \text{nonzero}: \mathbb{N} \times \mathbb{N} \to \text{Bool}$$

where nonzero: $\mathbb{N} \to$ Bool is the unary predicate with nonzero(x) = true if and only if $x \neq 0$. Remember that the semicolon denotes the composition of $\mathbb{B}$-functions.

Starting with the argument tuple (15,9) the $\mathbb{B}$-function (*) over NAT produces the following sequence of intermediate results:

$$(15,9) \mapsto (9,6) \mapsto (6,3) \mapsto (3,0) \mapsto 3.$$

In the preceding looping construction the else-part is of no interest, because the source tupling of the first and second projections forms the identity of pairs of natural numbers. Usually one allows to omit the else-part of a loop expression if its arguments represent the identity. With this convention the factorial function $n!$ can be expressed as follows:

$$(\text{id}_{\mathbb{N}}, \text{id}_{\mathbb{N}}); \quad \text{while } p_2 = 0 \quad \text{do} \quad (\text{times}, p_2; \text{pred}); p_2$$

where pred: $\mathbb{N} \circ\!\!\to \mathbb{N}$ denotes the unary partial predecessor function and times: $\mathbb{N} \times \mathbb{N} \to \mathbb{N}$ denotes the multiplication.

Even if we used some notations for the preceding examples of $\mathbb{B}$-functions over NAT we will not define any specific syntax for the functional language but we will define an abstract syntax that contains the context conditions of the single constructions. The following specification SIGNA-

TURE describes the knowledge on the base canon that is used within the abstract syntax.

> SIGNATURE **is** BOOL **with requirement**
>   **sorts** Sorts, Operators
>   **oprn** bool → Sorts
>   **with definition**
>   **sorts** Sorts*
>   **oprn** nil → Sorts*
>   app(Sorts*,Sorts) → Sorts*
>   **with requirement**
>   **oprn** source(Operators) → Sorts*
>   target(Operators) → Sorts
> **end** SIGNATURE

The following enrichment by definition of SIGNATURE represents the parameterized abstract syntax of the functional language.

> FUNCTIONS **is** SIGNATURE **with** NAT **with definition**
>   **oprn** lenght(Sorts*) → Nat
>   concat(Sorts*,Sorts*) → Sorts*
>   val(w:Sorts*,i:Nat **iff** $i \leq$ length(w)) → Sorts
>   **axioms** s: Sorts, w,w_1,w_2: Sorts*
>   length(nil) = zero
>   length(app(w,s)) = succ(length(w))
>   contact(w,nil) = w
>   concat(w_1,app(w_2,s)) = app(concat(w_1,w_2),s)
>   **if** w = concat(app(w_1,s),w_2), length(app(w_1,s))=i **then**
>   val(w,i) = s **fi**
>   **with definition**
>   **sorts** F, F*
>   **oprn** in(F) → Sorts*
>   out(F) → Sorts*
>   base(Operators) → F
>   compose(f:F, g:F, **iff** out(f)=in(g)) → F
>   prj(w:Sorts*, i:Nat **iff** $i \leq$ length(w)) → F
>   loop(p:F, g:F, f:F **iff** in(p)=in(g)=in(f)=out(f)=
>   =out(g), out(p)=app(nil,bool)) → F
>   branch(p:F, g:F, f:F **iff** in(p)=in(g)=in(f),
>   out(f)=out(g),out(p)=app(nil,pool)) → F
>   away(Sorts*) → F
>   first(F) → F*
>   add(f*:F*, f:F **iff** in*(f*)=in(f)) → F*
>   in*(F*) → Sorts*
>   tuple(F*) → F

axioms x: Operators, p,f,g,h: F, g^*, f^*: F^*, i: Nat, w: Sorts*
in(base(x)) = source(x)
out(base(x)) = app(nil,target(x))
if h = compose(f,g) **then** in(h) = in(f)
if h = compose(f,g) **then** out(h) = out(g)
if f = prj(w,i) **then** in(f) = w
if f = prj(w,i) **then** out(f) = app(nil,val(w,i))
if h = loop(p,f,g) **then** in(h) = in(f)
if h = loop(p,f,g) **then** out(h) = in(h)
if h = branch(p,f,g) **then** out(h) = out(f)
if h = branch(p,f,g) **then** in(h) = in(f)
in(away(w)) = w
out(away(w)) = nil
in*(first(f)) = in(f)
if g^* = add(f^*,f) **then** in*(g*) = in*(f^*)
in(tuple(f^*)) = in*(f^*)
out(tuple(first(f))) = out(f)
if g^* = add(f^*,f) **then**
out(tuple(g^*)) = concat(out(tuple(f^*),out(f)) **fi**

end FUNCTIONS

This specification needs some explanation. In the first enrichment by definition some auxiliary operations on sequences of elements of 'Sorts' are specified. The operator 'length(Sorts*) → Nat' gives the length for each $w \in$ Sorts*, and 'concat(Sorts*,Sorts*) → Sorts*' represents the concatenation of sequences of sorts, and 'val(Sorts*,Nat) ⤏ Sorts yields the i-th element of a sequence. The second enrichment by definition also contains some auxiliary operations and the auxiliary sort F^* which represents finite non-empty sequences of **B**-functions such that all **B**-functions in a sequence have equal inputs whereas **B**-functions contained in different sequences may have different inputs. The operators 'in(F) → Sorts*' and 'out(F) → Sorts*' associate with each **B**-function its input and output sequence of sorts. These operators are necessary for the context conditions of the single constructions of **B**-functions. These constructions are represented by the operators 'base', 'compose', 'prj', 'loop', 'branch', 'away', and 'tuple'.

Besides the constructions mentioned in the informal presentation the formal specification contains the construction 'base(Operator) → F' that transforms each operator into a **B**-function. The operator 'away(Sorts*) → F' represents the source tupling of empty sequences of **B**-functions, and the operator 'tuple(F^*) → F' represents the source tupling of non-empty sequences of **B**-functions.

Now we are able to define the algebra A that defines the semantics of the functional language.

First of all we set

$$A_{\text{Sorts}} = \text{sorts}(\mathbb{B}), \qquad A_{\text{Operators}} = \text{operators}(\mathbb{B}),$$

$$\text{source}^A = \text{source}^{\mathbb{B}}, \qquad \text{target}^A = \text{target}^{\mathbb{B}},$$

$$A_{\text{Nat}} = \mathbb{N}, \qquad A_{\text{Bool}} = \{\text{true,false}\}, \qquad A_{\text{Sorts}\,*} = \text{sorts}(\mathbb{B})^*,$$

and nil^A, app^A, length^A, concat^A, and val^A are interpreted as the empty sequence of sort names of $\mathbb{B}$, as the appending of a sort name to a given sequence of sort names, as the length of sequences of sort names, as the concatenation of sequences of sort names, and as the selection of the i-th component of a sequence, respectively.

The interpretation of the remaining sorts F and F^* and of the remaining operators is the kernel of the semantics definition.

We start with the definition of the sets A_F and A_{F*}:

$A_F = \{(\hat{\mathbb{B}}, \boldsymbol{t}: w \multimap u) \mid \hat{\mathbb{B}}$ is an enrichment by definition of $\mathbb{B}$ and $\boldsymbol{t}: w \multimap u$ with $w, u \in \text{sorts}(\mathbb{B})^*$ is a term sequence on $\hat{\mathbb{B}}\}$;

$A_{F*} = \{(\hat{\mathbb{B}}, \boldsymbol{t}^{\#}: w \multimap u_1 \# u_2 \# \ldots \# u_r) \mid \hat{\mathbb{B}}$ is an enrichment by definition of $\mathbb{B}$ and the second component is a $\mathbb{B}$-function sequence with $w, u_1, \ldots, u_r \in \text{sorts}(\mathbb{B})^*\}$

The interpretations of the operators 'in', 'out' and 'in*' will be defined as expected:

$$\text{in}^A(\hat{\mathbb{B}}, \boldsymbol{t}: w \multimap u) = w,$$

$$\text{out}^A(\hat{\mathbb{B}}, \boldsymbol{t}: w \multimap u) = u,$$

$$(\text{in}^*)^A\,(\hat{\mathbb{B}}, \boldsymbol{t}^{\#}: w \multimap u_1 \# u_2 \# \ldots \# u_r) = w.$$

Since the sort F^* represents only an auxiliary construction of sequences of term sequences the interpretations of the operators 'first' and 'add' are naturally given by

$$\text{first}^A(\hat{\mathbb{B}}, \boldsymbol{t}: w \multimap u) = (\hat{\mathbb{B}}, \boldsymbol{t}^{\#}: w \multimap u)$$

where $\boldsymbol{t}: w \multimap u$ is interpreted as a sequence of length one, and

$$\text{add}^A\big((\hat{\mathbb{B}}_1, t_1^{\#}: w \multimap u_1 \# u_2 \# \ldots \# u_r), (\mathbb{B}_2, \boldsymbol{t}: w \multimap u)\big)$$
$$= (\hat{\mathbb{B}}_1 \textbf{ with } \hat{\mathbb{B}}_2, t_2^{\#}: w \multimap u_1 \# u_2 \# \ldots \# u_r \# u)$$

such that $\text{length}(\boldsymbol{t}_2^{\#}) = \text{length}(\boldsymbol{t}_1^{\#}) + 1$, and $t_2^{\#}(i) = t_1^{\#}(i)$ for $1 \leq i \leq r$ and $\boldsymbol{t}_2^{\#}(r + 1) = \boldsymbol{t}$.

The most interesting part of the definition of the semantics is the interpretation of the operators 'base', 'compose', 'prj', 'loop', 'branch', 'away', and 'tuple' since these operators represent the semantically meaningful constructions of $\mathbb{B}$-functions from predefined $\mathbb{B}$-functions.

Semantics of 'base'

For each operator $\sigma: w \circ\!\!\rightarrow S$ in the signature of the initial canon $\mathbb{B}$ we define

$$\text{base}^A(\sigma) = (\mathbb{B}, \sigma: w \circ\!\!\rightarrow S)$$

where the operator is considered as a term over $\mathbb{B}$, and that term is considered as a sequence of one component.

Semantics of 'compose'

The elements $(\hat{\mathbb{B}}_1, \boldsymbol{t}_1: w_1 \circ\!\!\rightarrow u_1)$, $(\hat{\mathbb{B}}_2, \boldsymbol{t}_2: w_2 \circ\!\!\rightarrow u_2)$ of A_F are in the domain of the partial composition 'compose^A' if and only if $u_1 = w_2$. In that case we set

$$\begin{aligned} &\text{compose}^A\big((\hat{\mathbb{B}}_1, \boldsymbol{t}_1: w_1 \circ\!\!\rightarrow u_1), (\hat{\mathbb{B}}_2, \boldsymbol{t}_2: u_1 \circ\!\!\rightarrow u_2)\big) \\ &\qquad = (\hat{\mathbb{B}}_1 \ \textbf{with}\ \hat{\mathbb{B}}_2, \boldsymbol{t}: w_1 \circ\!\!\rightarrow u_2) \end{aligned}$$

with $\boldsymbol{t}(i) = [\boldsymbol{t}_2(i)]\,(\boldsymbol{t}_1)$ for $1 \le i \le \text{length}(u_2)$, where

$$[\boldsymbol{t}_2(i)]\,(\boldsymbol{t}_1): w_1 \circ\!\!\rightarrow u_2(i)$$

denotes the term that results from $\boldsymbol{t}_2(i): u_1 \circ\!\!\rightarrow u_2(i)$ by the simultaneous substitution of $\boldsymbol{t}_1(j)$ for the j-th variable in $\boldsymbol{t}_2(i): u_1 \circ\!\!\rightarrow u_2(i)$ for each j with $1 < j < \text{length}(u_1)$.

We demonstrate this composition of term sequences by a simple example. Let

$$\boldsymbol{t}_1 = (\text{g.c.d}: \text{Nat} \times \text{Nat} \rightarrow \text{Nat}, \text{plus}: \text{Nat} \times \text{Nat} \rightarrow \text{Nat})$$

and

$$\begin{aligned} \boldsymbol{t}_2 = (&\text{less}: \text{Nat} \times \text{Nat} \rightarrow \text{Bool}, \text{times}: \text{Nat} \times \text{Nat} \rightarrow \text{Nat}, \\ &\text{l.c.m}: \text{Nat} \times \text{Nat} \rightarrow \text{Nat}) \end{aligned}$$

be (NAT **with** BOOL)-functions. The composition of $\boldsymbol{t}_1$ with $\boldsymbol{t}_2$ yields

$$\boldsymbol{t}: \text{Nat} \times \text{Nat} \rightarrow \text{Bool} \times \text{Nat} \times \text{Nat}$$

that produces a triple

$$(y_1, y_2, y_3) \in \{\text{false}, \text{true}\} \times \mathbb{N} \times \mathbb{N}$$

for each pair $(x,y) \in \mathbb{N} \times \mathbb{N}$ in such a way that

$$\begin{aligned} y_1 &= \text{g.c.d}(x,y) < (x + y) \quad (= \text{true for all } x,y), \\ y_2 &= \text{g.c.d}(x,y) \cdot (x + y), \\ y_3 &= \text{l.c.m}(\text{g.c.d}(x,y), x+y). \end{aligned}$$

Semantics of 'prj'

In the specifications of the preceding sections projections that pick out some component of an n-tuple of values do not explicitly occur. Instead of those projections we used variables, and the evaluation of a variable is the equivalent to the application of a projection. However, in the examples within the informal presentation of the functional language we used explicit projections. We illustrate the equivalence between the use of variables and the use of projections by a simple example. Let us consider a declaration of the form

$$x,y\colon \text{Nat},\quad r\colon \text{String},\quad a,b\colon \text{Bool}$$

then the variable r represents the third projection

$$p_3\colon \text{Nat}*\text{Nat}*\text{String}*\text{Bool}*\text{Bool} \to \text{String}$$

whereas the variable y represents the projection

$$p_2\colon \text{Nat}*\text{Nat}*\text{String}*\text{Bool}*\text{Bool} \to \text{Nat}$$

that picks out the second element of a given 5-tuple.

For the definition of the semantics of the functional language we use projections. For each sequence w of $\mathbb{B}$-types and each natural number i with $1 \le i \le \text{length}(w)$

$$p_i^w\colon w \multimap w(i)$$

denotes the i-th projection. With this convention we can define

$$\text{prj}^A(w,i) = \big(\mathbb{B},\, p_i^w\colon w \multimap w(i)\big).$$

Semantics of 'away'

The intended semantics of this construction was the $\mathbb{B}$-function that takes away each argument. In the framework of $\mathbb{B}$-function sequences this construction can be represented by the empty $\mathbb{B}$-function sequence that will be denoted by

$$\emptyset\colon w \multimap \text{nil}^A$$

where nil^A denotes the empty sequence of sort names. With this notation we can define

$$\text{away}^A(w) = (\mathbb{B}, \emptyset\colon w \multimap \text{nil}^A).$$

Semantics of 'tuple'

Up to this point the defined constructions generate only $\mathbb{B}$-functions that produce single values, i.e. sequences of length one. The construction that generates $\mathbb{B}$-functions is the source tupling represented by the operator 'tuple'. In the framework of $\mathbb{B}$-function sequences this construction transforms a $\mathbb{B}$-function sequence into one $\mathbb{B}$-function. The formal definition is as follows:

$$\text{tuple}^A(\hat{\mathbb{B}}, \boldsymbol{t}^{\#}: w \multimap u_1 \# u_2 \# \ldots \# u_r) = (\hat{\mathbb{B}}, \boldsymbol{t}: w \multimap u)$$

where $u = u_1 * u_2 * \ldots * u_r \in \text{sorts}(\mathbb{B})^*$ is the concatenation of $u_1, u_2, \ldots, u_r$ and

$$\boldsymbol{t}(k) = \big(\boldsymbol{t}^{\#}(i)\big)(j) \text{ if } k = \text{length}(u_1) + \ldots + \text{length}(u_{i-1}) + j \text{ and}$$
$$\text{length}(u_{i-1}) < k \leq \text{length}(u_i),$$
$$\boldsymbol{t}(k) = \big(\boldsymbol{t}^{\#}(1)\big)(k) \quad \text{if} \quad 1 \leq k \leq \text{length}(u_1).$$

Semantics of 'loop'

The semantics of the loop-construction will be defined by a suitable modification of the previously defined initial canon WHILE.

Let

$$(\hat{\mathbb{B}}_1, t_1: w \multimap \text{Bool}), \qquad (\hat{\mathbb{B}}_2, \boldsymbol{t}_2: w \multimap w), \qquad (\hat{\mathbb{B}}_3, \boldsymbol{t}_3: w \multimap w)$$

be given arguments of the partial operation 'loop^A' contained in the domain of that partial operation. Then the intended semantics requires the substitution of the parameter sort M in the initial canon WHILE by the sequence $w \in \text{sorts}(\mathbb{B})^*$, and of the parameter operators

$$p(M) \multimap \text{Bool}, \quad f(M) \multimap M, \quad g(M) \multimap M$$

by

$$t_1: w \multimap \text{Bool}, \quad \boldsymbol{t}_2: w \multimap w, \quad \boldsymbol{t}_3: w \multimap w$$

respectively.

This kind of modification requires a slight generalization of the modification of initial canons as defined up to now. On the left-hand side of a modification list we now permit sequences of sort names and $\mathbb{B}$-function sequences of the modified canon. With this generalization we are able to define

$\text{loop}^A((\hat{\mathbb{B}}_1, t_1: w \multimap \text{Bool}), (\hat{\mathbb{B}}_2, \boldsymbol{t}_2: w \multimap w), (\hat{\mathbb{B}}_3, \boldsymbol{t}_3: w \multimap w)) =$
$= ($WHILE **modiefied by** ($\hat{\mathbb{B}}_1$ **with** $\hat{\mathbb{B}}_2$ **with** $\hat{\mathbb{B}}_3$) **according**
w **for** M
$\boldsymbol{t}_1$ **for** p
$\boldsymbol{t}_2$ **for** f
$\boldsymbol{t}_3$ **for** g
while $\boldsymbol{t}_1$ do $\boldsymbol{t}_3$ else $\boldsymbol{t}_2$ **for** while **end,**
while $\boldsymbol{t}_1$ do $\boldsymbol{t}_3$ else $\boldsymbol{t}_2: w \multimap w$)

If the semantics of the loop-construction is given by the following flow-chart, then this change in the semantics of the loop-construction can easily be realized by the use of a corresponding initial canon LOOP defined in the following.

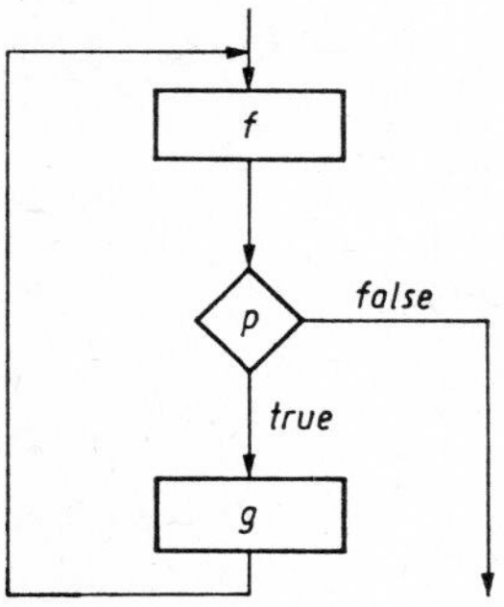

The kernel of the changed definition of the semantics is the specification of the initial canon LOOP.

LOOP is BOOL **with requirement**
  sorts M
  oprn $p(M) \multimap$ Bool
    $f(M) \multimap M$
    $g(M) \multimap M$
  with definition
  oprn loop(M) $\multimap M$
  axioms $x: M$
  if $p(f(x)) =$ false **then** loop(x) $= f(x)$
  if $p(f(x)) =$ true, loop($g(f(x))$) $=$ loop($g(f(x))$) **then**
    loop(x) $=$ loop($g(f(x))$) **fi**
end LOOP

Using this specification we define

loop^A(($\hat{\mathbb{B}}_1$, $\boldsymbol{t}_1: w \multimap$ Bool),($\hat{\mathbb{B}}_2$, $\boldsymbol{t}_2: w \multimap w$),($\hat{\mathbb{B}}_3$, $\boldsymbol{t}_3: w \multimap w$)) $=$
  $=$ (LOOP **modified by** ($\hat{\mathbb{B}}_1$ **with** $\mathbb{B}_2$ **with** $\hat{\mathbb{B}}_3$) **according**
    w **for** M
    $\boldsymbol{t}_1$ **for** p
    $\boldsymbol{t}_2$ **for** f
    $\boldsymbol{t}_3$ **for** g
    loop($\boldsymbol{t}_1$,$\boldsymbol{t}_2$,$\boldsymbol{t}_3$) **for** loop **end,**
  loop($\boldsymbol{t}_1$,$\boldsymbol{t}_2$,$\boldsymbol{t}_3$)$: w \multimap w$)

Semantics of 'branch'

The semantics of the branch-construction will be defined in the same way as the semantics of the loop-construction. Evidently, we have to use another initial canon. We start with the specification of the initial canon BRANCH.

> **BRANCH is** BOOL **with requirement**
>   **sorts** P,M
>   **oprn** $p(P)$ ⟴ Bool
> $f(P)$ ⟴ M
> $g(P)$ ⟴ M
>   **with definition**
>   **oprn** branch(P) ⟴ M
>   **axioms** $x: P$
>   **if** $p(x)$ = true, $f(x) = f(x)$ **then** branch(x) = $f(x)$
>   **if** $p(x)$ = false, $g(x) = g(x)$ **then** branch(x) = $g(x)$
> **end** BRANCH

Using this specification we define

> branch^A(($\hat{\mathbb{B}}_1$, $\boldsymbol{t}_1$: w ⟴ Bool), ($\hat{\mathbb{B}}_2$, $\boldsymbol{t}_2$: w ⟴ u), ($\hat{\mathbb{B}}_3$, $\boldsymbol{t}_3$: w ⟴ u)) =
>   = (BRANCH **modified by** ($\hat{\mathbb{B}}_1$ **with** $\hat{\mathbb{B}}_2$ **with** $\hat{\mathbb{B}}_3$) **according**
> w **for** P
> u **for** M
> $\boldsymbol{t}_1$ **for** p
> $\boldsymbol{t}_2$ **for** f
> $\boldsymbol{t}_3$ **for** g
>   if $\boldsymbol{t}_1$ then $\boldsymbol{t}_2$ else $\boldsymbol{t}_3$ **for** branch **end**,
>   if $\boldsymbol{t}_1$ then $\boldsymbol{t}_2$ else $\boldsymbol{t}_3$: w ⟴ $\boldsymbol{u}$).

The preceding semantics of the branching does not assume the applicability of both alternatives. Therefore, a further conceivable semantics of the branching could be given by the following specification BRANCH* defined as follows.

> **BRANCH* is** BOOL **with requirement**
>   **sorts** P,M
>   **oprn** $p(P)$ ⟴ Bool
> $f(P)$ ⟴ M
> $g(P)$ ⟴ M
>   **with definition**
>   **oprn** branch*(P) ⟴ M
>   **axioms** $x: P$
>   **if** $p(x)$ = false, $f(x)=f(x)$, $g(x)=g(x)$ **then** branch*(x)=$g(x)$
>   **if** $p(x)$=true, $f(x)=f(x)$, $g(x)=g(x)$ **then** branch*(x)=$f(x)$
> **end** BRANCH*

The initial canon BRANCH* presumes the applicability of both alternatives and of the branching condition as a sufficient domain condition of 'branch*(P) $\circ\!\!\rightarrow M$', and due to the initiality of the last enrichment in BRANCH* this condition is also a necessary domain condition.

The homomorphism $\Phi: I \rightarrow A$, uniquely defined by the semantics algebra A, is not surjective, i.e. not each ($\hat{\mathbb{B}}, \boldsymbol{t}: w \circ\!\!\rightarrow u$) is the meaning of an abstract $\mathbb{B}$-function. If for instance $\hat{\mathbb{B}}$ contains an enrichment by definition of $\mathbb{B}$ that introduces a new sort name, then that pair cannot be the meaning of an abstract $\mathbb{B}$-function, since the defined language does not contain facilities for the definition of new data types.

Finally we repeat that it is not the aim of this section to introduce a functional language and to contribute in this way to the research work on functional programming; rather it is to demonstrate the applicability of structural induction even when the generating operations are not defined for all arguments and to use for this demonstration a non-trivial example.

2. Equational partiality

Whereas in the first chapter it is shown that there are strong motivations for the development of a theory of equationally partial many-sorted algebras, the following chapter develops the algebraic basis of the preceding use of structural induction on partial algebras. According to that aim the following two chapters are a contribution to the universal algebra of partial algebraic structures.

The existence of different sorts of elements in all examples of Chapter 1 is algebraically reflected by the use of many-sorted algebras (or heterogeneous algebras as they are called by BIRKHOFF and LIPSON (1970)). However, the use of many-sorted algebras causes some notational difficulties. These notational problems could be considerably reduced if we were to use the mathematical language of category theory, but this language is a very abstract one and the uses of abstraction to achieve generality are not the same as in other branches of mathematics and computer science. The theory of categories is a very useful and powerful tool in the hands of a well-trained mathematician. Because most mathematicians and specialists in computer science are not familiar with the basic concepts of category theory, we will not use this tool here. Readers who are interested in category theory should consult the famous book *Categories for the working mathematician* by MACLANE (1971) or a recent paper of GOGUEN and BURSTALL (1983). In the following we will use the traditional language of sets, functions, operations, relations, and so on. We deal mostly with S-indexed families and functions between S-indexed families of sets, where S represents the set of different sorts of objects.

To overcome the notational problems of many-sorted algebras we introduce in the first section the underlying set-theoretical notions and concepts. In the second section we introduce the concepts of total many-sorted algebras, of partial many-sorted algebras, of homomorphisms and of so-called existence equations. The third section is devoted to defining the concept of equational partiality and equationally partial algebras. The fourth section deals with hierarchy conditions for both system of domain conditions of operators and sets of axioms. It is proved that hierarchical systems of domain conditions are necessary and sufficient for the definition of substructures by the requirement that the inclusion forms a homomorphism. This implies that for this theory only one concept of homomorphism is necessary to yield a nice theory of structure.

On the basis of the results of the fourth section we introduce in the last section of this chapter hierarchical equationally partial varieties, briefly hep-varieties, and hep-quasi-varieties. The class of small categories turns out to be a typical example of a hep-variety.

Concerning the use of partial algebras in data abstraction we refer to BERGSTRA et al. (1981). They assess that partial functions occur everywhere in program specification but not so much work has been done in this field. Beside the equational partiality that will be developed in the following chapters, M. BROY and M. WIRSING also use partial algebras in data and program specifications, see BROY and WIRSING (1981), (1982). In the papers of 1981 they use the concept of existence equations as developed in REICHEL (1979) and ANDREKA, BURMEISTER, NEMETI (1980). This approach will be thoroughly developed in the following two chapters. In the paper of 1982 BROY and WIRSING use the so-called *strong equality* of equations, which says that a given set of values of the variables in an equation $t = t'$ constitutes a solution of that equation iff both sides are undefined or both sides are defined and produce the same values. This strong equality implies the use of an additional semantic predicate $D(t)$ which states that the term is defined, i.e. for an algebra A the interpretation t^A denotes some object of the algebra A. The term t is assumed to be without variables. If one mainly deals with finitely generated partial algebras, i.e. partial algebras without proper subalgebras, then the differences between this and the use of existence equations are not very great. The differences and the reasons for generally preferring existence equations are discussed in BURMEISTER (1986).

Recently, KAMIN and ARCHER (1984) used the concept of *partial algebras with preconditions* for the notion of partial implementations of abstract data types. This approach seems to be an interesting alternative to error values in specifications that are mainly based upon implementation considerations. They suggest that error values should not be used in abstract specifications but to allow partial approximations of the specified total algebras as correct representations. This idea does not need a thoroughly developed theory of partial algebras and the paper contains only some basic notions on this point.

2.1. Set-theoretical notions

As usual we denote sets by capital letters, elements and functions by small letters. For finite sets the elements which belong to the set can also be given within braces.

Let S be any finite set. An S-indexed family $A = (A_s \mid s \in S)$ of sets will briefly be called an *S-set*. Correspondingly, an S-indexed family $f = (f_s\colon A_s \to B_s \mid s \in S)$ of functions will briefly be called an *S-function* or *S-mapping*. If it is evident from context that both A and B are S-sets, we denote an S-function by $f\colon A \to B$. For any $s \in S$ the set A_s and the mapping $f_s\colon A_s \to B_s$ will be called the *s-component* of the S-set A and of the S-mapping $f\colon A \to B$ respectively.

Let $f\colon A \to B$ be any function and $x \in A$ any element of the domain A of $f\colon A \to B$. The element that is assigned to the element x by the function $f\colon A \to B$ will be denoted by $xf \in B$. If additionally $g\colon B \to C$ is a further mapping, the composition will be denoted by $fg\colon A \to C$, sometimes also by $f \circ g\colon A \to C$. The application of the composition to any $x \in A$ is then denoted by

$$xfg \in C \quad \text{or by} \quad (xf)\,g \in C$$

if we want to stress the successive application of f and g. This notation is compatible both with the diagrammatic and the applicative order of the composition. We hope that readers will not be confused or irritated by the use of xf instead of $f(x)$.

Very often we have to deal with sets of variables taking values in an S-set $A = (A_s \mid s \in S)$. In this case the particular s-component in which each variable $x \in X$ can take on values must be fixed. Accordingly we define an *S-sorted system of variables* as a mapping $v\colon X \to S$. Now, any $x \in X$ can take on values in A_{xv} only. An assignment of $v\colon X \to S$ in an S-set $A = (A_s \mid s \in S)$ is therefore given by a function

$$\boldsymbol{a}\colon X \to \bigcup_{s \in S} A_s$$

such that $x\boldsymbol{a} \in A_{xv}$ for each $x \in X$. A_v denotes the set of all *assignments* of the S-sorted system of variables $v\colon X \to S$ in the S-set $A = (A_s \mid s \in S)$.

Let $f = (f_s\colon A_s \to B_s \mid s \in S)$ be any S-function and $v\colon X \to S$ any S-sorted system of variables, then we can extend in a natural way the S-function to a function

$$f_v\colon A_v \to B_v$$

by setting $x(\boldsymbol{a}f_v) = (x\boldsymbol{a})\,f_{xv} \in B_{xv}$ for every $x \in X$, $\boldsymbol{a} \in A_v$. The mapping $f_v\colon A_v \to B_v$ transforms any assignment $\boldsymbol{a} \in A_v$ to an assignment $\boldsymbol{a}f_v \in B_v$ of the same S-sorted system of variables $v\colon X \to S$ in the S-set $B = (B_s \mid s \in S)$ by pointwise application of the S-function $f\colon A \to B$.

For the sake of simpler formula we allow the subscript s and v to be omitted from the functions $f_s\colon A_s \to B_s$, $f_v\colon A_v \to B_v$ respectively, if this subscript can easily be derived from context.

$\mathbb{N} = \{0, 1, 2, \ldots\}$ denotes the set of natural numbers, and for any $n \in \mathbb{N}$ we set $[n] = \{1, 2, \ldots, n\}$ and $[0] = \emptyset$. A finite S-sorted system of variables $w\colon [n] \to S$ can obviously be identified with the string $w = s_1 s_2 \ldots s_n$ where $s_i = iw$ for every $i \in [n]$. The empty string will be denoted as usual by $\lambda\colon [0] \to S$. S^* denotes the set of all finite strings of elements of S, e.g. the set of all S-sorted systems of variables of the form $w\colon [n] \to S$. S^+ denotes the set of all non-empty strings of elements of S.

For any string $w\colon [n] \to S$ an assignment $\boldsymbol{a} \in A_w$ in an S-set $A = (A_s \mid s \in S)$ will also be denoted by

$$\boldsymbol{a} = (a_1, a_2, \ldots, a_n) \quad \text{with} \quad a_i = i\boldsymbol{a} \quad \text{for every} \quad i \in [n]$$

as ordered n-tuples are usually denoted. According to that convention A_w can be identified with the Cartesian product, i.e.

$$A_w = A_{1w} \times A_{2w} \times \ldots \times A_{nw}.$$

We recall that for the empty string $\lambda\colon [0] \to S$ the set A_λ consists of one element for every S-set $A = (A_s \mid s \in S)$ which is the inclusion of the empty set into the union of the s-components of the S-set. Therefore we can set $A_\lambda = \{\emptyset\}$.

If $w\colon [n] \to S$ is a string, $A = (A_s \mid s \in S)$ is an S-set, $f\colon A_w \to M$ is any function, and if we use the notation $(a_1, \ldots, a_n)$ to denote elements of A_w, then the value associated with $(a_1, \ldots, a_n)$ by $f\colon A_w \to M$ will be denoted by $(a_1, \ldots, a_n)\, f$.

A binary S-relation ϱ in an S-set $A = (A_s \mid s \in S)$ is given by an S-indexed family $\varrho = (\varrho_s \mid s \in S)$ such that $\varrho_s \subseteq A_s \times A_s$ for every $s \in S$. An S-relation ϱ in an S-set A is referred to as an S-equivalence in A if the binary relation $\varrho_s \subseteq A_s \times A_s$ is an equivalence relation for every $s \in S$, i.e. if ϱ_s is reflexive, transitive and symmetric for every $s \in S$.

Finally we will point out an important effect. Let $v\colon X \to S$ be a sorted system of variables and $A = (A_s \mid s \in S)$ an S-set so that at least one variable $x \in X$ exists with $A_{xv} = \emptyset$. Then there exists no assignment of $v\colon X \to S$ in A so $A_v = \emptyset$. This effect causes the theory of many-sorted algebras to differ seriously from the theory of homogeneous algebras, especially with respect to the *equational theory* of many-sorted algebras.

2.2. Many-sorted algebras

Before we deal with partial algebras we introduce the basic notions of total many-sorted algebras.

Definition 2.2.1. A *signature*

$$\Sigma = (S, \alpha\colon \Omega \to S^* \times S)$$

is given by a finite set S of so-called *sort names*, by a finite set Ω of operators or operation symbols, and by an arity function $\alpha\colon \Omega \to S^* \times S$ which associates with every operator $\sigma \in \Omega$ an ordered pair $\sigma\alpha = (w, s)$. The string w is called the *domain of* σ and $s \in S$ the *range of* $\sigma \in \Omega$, respectively. Instead of $\sigma\alpha = (w, s)$ we often write $\sigma\colon w \to s$.

A Σ-*algebra*

$$A = \big((A_s \mid s \in S),\, (\sigma^A \mid \sigma \in \Omega)\big)$$

is given by an S-set $(A_s \mid s \in S)$, the underlying S-set of A, and by an Ω-indexed family $(\sigma^A \mid \sigma \in \Omega)$ of fundamental operations, so that for every operator $\sigma \in \Omega$ with $\sigma\colon w \to s$ the associated fundamental operation σ^A is a total function from A_w to A_s, i.e.

$$\sigma^A: A_w \to A_s.$$

For a Σ-algebra A the underlying S-set of A will also be denoted by A and will be called the *S-carrier* of the Σ-algebra A. By $\mathrm{ALG}_t(\Sigma)$ we denote the class of all Σ-algebras.

If $A, B \in \mathrm{ALG}_t(\Sigma)$, an S-function $f = (f_s: A_s \to B_s \mid s \in S)$ is said to be a *Σ-homomorphism from A to B*, denoted by $f: A \to B$, if for every $\sigma \in \Omega$ with $\sigma: w \to s$ and every $\boldsymbol{a} \in A_w$ the following equation holds:

$$(\boldsymbol{a}\sigma^A)\, f_s = (\boldsymbol{a} f_w)\, \sigma^B,$$

i.e. the following diagram commutes:

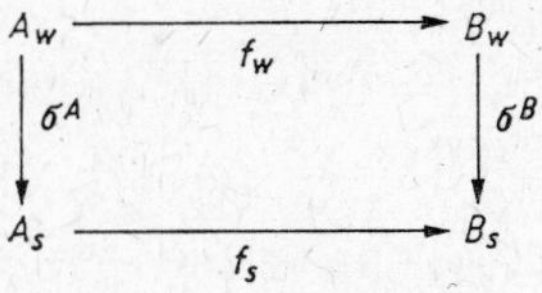

for every $\sigma \in \Omega$.

A Σ-homomorphism $f: A \to B$ is called a *Σ-isomorphism,* if there is a Σ-homomorphism $g: B \to A$ with

$$f_s \cdot g_s = \mathrm{Id}_{A_s}, \qquad g_s \cdot f_s = \mathrm{Id}_{B_s} \quad \text{for every} \quad s \in S. \qquad \blacksquare$$

When we give examples of signatures we will use a more readable representation similar to the specifications of the first chapter.

In the application of algebra in computer science signatures with additional structure serve as mathematical representations of abstract data types, see Chapters 1 and 4. Relations between abstract data types and constructions with them will be mathematically described by relations and constructions on structured signatures. For this reason we introduce the notion of a subsignature and of a signature morphism.

Definition 2.2.2. $\Sigma_1 = (S_1, \alpha_1: \Omega_1 \to S_1^* \times S_1)$ is a *subsignature of* $\Sigma_2 = (S_2, \alpha_2: \Omega_2 \to S_2^* \times S_2)$, written $\Sigma_1 \subseteq \Sigma_2$, if $S_1 \subseteq S_2$, $\Omega_1 \subseteq \Omega_2$, and $\sigma\alpha_1 = \sigma\alpha_2$ for every $\sigma \in \Omega_1$. A *signature morphism* $(\varphi, \psi): \Sigma_1 \to \Sigma_2$ is an ordered pair (φ, ψ), where $\varphi: S_1 \to S_2$, $\psi: \Omega_2 \to \Omega_2$ such that $\psi \cdot \alpha_2 = \alpha_1 \cdot (\varphi^* \times \varphi)$, where $(w, s)(\varphi^* \times \varphi) = (w \cdot \varphi, s\varphi)$ for every $w: [n] \to S_1$, $s \in S_1$. The string $w \cdot \varphi \in S_2^*$ results from $w: [n] \to S_1$ by pointwise application of $\varphi: S_1 \to S_2$, i.e. $i(w \cdot \varphi) = (iw)\, \varphi$ for each $i \in [n]$.

The composition of signature morphism $(\varphi, \psi): \Sigma_1 \to \Sigma_2$ and $(\overline{\varphi} \cdot \overline{\psi})$: $\Sigma_2 \to \Sigma_3$ is naturally defined by

$$(\varphi, \psi) \cdot (\overline{\varphi}, \overline{\psi}) = (\varphi \cdot \overline{\varphi}, \psi \cdot \overline{\psi}),$$

i.e. by componentwise composition. $\blacksquare$

Clearly, $\Sigma_1 \subseteq \Sigma_2$ holds iff (if and only if) the inclusions $S_1 \subseteq S_2$ and $\Omega_1 \subseteq \Omega_2$ form a signature morphism.

We can omit examples of subsignatures and signature morphisms here because they can be found in Chapters 1 and 4.

At this point we will draw attention to a construction indispensable for almost all following notions and investigations. Roughly speaking we will make a closer study of all formal expressions which can be constructed from an S-sorted system of variables $v: X \to S$ and a signature Σ, and we will consider the external behaviour of this term algebra $T(\Sigma, v)$ with respect to total and partial Σ-algebras. We will see that this external behaviour, described by the following recursion theorem, characterizes $T(\Sigma, v)$ uniquely up to isomorphism. Therefore, it is not the special construction of $T(\Sigma, v)$ that is important but the external behaviour. We will construct $T(\Sigma, v)$ by means of the so-called *postfix Polish notation.*

Definition 2.2.3. Let $\Sigma = (S, \alpha: \Omega \to S^* \times S)$ be any signature and $v: X \to S$ an S-sorted system of variables.

$$T(\Sigma, v) = \big(T(\Sigma, v)_s \mid s \in S\big)$$

is the smallest S-set whose elements are strings from $X \cup \Omega$ (it is supposed that $X \cap \Omega = \emptyset$) satisfying the following conditions:

(1) $x \in T(\Sigma, v)_{xv}$ for every $x \in X$;

(2) $\sigma \in T(\Sigma, v)_s$ for every $\sigma \in \Omega$ and $s \in S$ with $\sigma\alpha = (\lambda, s)$;

(3) If $\sigma \in \Omega$ with $\sigma: w \to s$, $w: [n] \to S$, $n \neq 0$ and if $\boldsymbol{t} = (t_1, \ldots, t_n) \in T(\Sigma, v)_w$ then $t_1 t_2 \ldots t_n \sigma \in T(\Sigma, v)_s$.

In the sequel $t_1 t_2 \ldots t_n \sigma$ will also be denoted briefly by $\boldsymbol{t}\sigma$. By (2) and (3) we can define easily the fundamental operation

$$\sigma^{T(\Sigma,v)}: T(\Sigma, v)_w \to T(\Sigma, v)_s \quad \text{for every} \quad \sigma: w \to s \text{ in } \Omega.$$

Let $\sigma\alpha = (\lambda, s)$, then

$$\sigma^{T(\Sigma,v)} = \sigma \in T(\Sigma, v)_s$$

and if $\sigma: w \to s$, $w: [n] \to S$, $n \neq 0$, and $\boldsymbol{t} \in T(\Sigma, v)_w$ then

$$\boldsymbol{t}\sigma^{T(\Sigma,v)} = \boldsymbol{t}\sigma \in T(\Sigma, v)_s.$$

We denote by

$$T(\Sigma, v) = \big(\big(T(\Sigma, v)_s \mid s \in S\big), (\sigma^{T(\Sigma,v)} \mid \sigma \in \Omega)\big)$$

the so-called *algebra of Σ-terms on* $v: X \to S$ (briefly the *Σ-term-algebra* on v). ■

This representation of terms will be used only in formal mathematical considerations. When we give examples, we will use a more readable notation with infix notation for binary operators and functional notation with parentheses for other operators in the same way as in Chapter 1. If we use terms or term equations to describe properties of operations, then the usual notation is very suitable, but if terms become objects of mathematical investigation another notion is more suitable, at least from our point of view.

Now we can turn to the external behaviour of $T(\Sigma, v)$.

Theorem 2.2.4 (*Recursion Theorem*). *Let* $\Sigma = (S, \alpha\colon \Omega \to S^* \times S)$ *be any signature and* $v\colon X \to S$ *an S-sorted system of variables. For any Σ-algebra A and any assignment* $\boldsymbol{a} \in A_v$ *there exists exactly one homomorphism*

$$a\colon T(\Sigma, v) \to A \quad \textit{with} \quad xa_{xv} = x\boldsymbol{a} \quad \textit{for every} \quad x \in X,$$

i.e. every assignment can uniquely be extended to a homomorphism.

Proof: The uniqueness of the extension to a homomorphism is evident, because $T(\Sigma, v)$ is generated by the S-set $\big(v^{-1}(s) \mid s \in S\big)$, where $v^{-1}(s) = \{x \in X \mid xv = s\}$, and because any two homomorphisms which coincide on a generating set are equal. So it remains to prove that $\boldsymbol{a} \in A_v$ can be extended to a homomorphism $a\colon T(\Sigma, v) \to A$. By appealing directly to the inductive definition of the S-set of Σ-terms on $v\colon X \to S$ we can define the homomorphism by $xa_{xv} = x\boldsymbol{a}$ for every $x \in X$, $\sigma a_s = \sigma^A$ for every $\sigma \in \Omega$, $s \in S$ with $\sigma\colon \lambda \to s$, and $(\boldsymbol{t}\sigma)\, a_s = (\boldsymbol{t}a_w)\, \sigma^A$ for every $\sigma \in \Omega$, $s \in S$ with $\sigma\colon w \to s$, $w\colon [n] \to S$, $n \neq 0$, and every $\boldsymbol{t} \in T(\Sigma, v)_w$.

Obviously, the S-function

$$a = \big(a_s\colon T(\Sigma, v)_s \to A_s \mid s \in S\big)$$

is uniquely defined in this way, and it is a homomorphism. ■

We illustrate the recursion theorem by a simple example: Let us consider the signature

Σ **is sorts** B, C, E
  oprn $f, t \to B$
    $0, 1 \to E$
    $+, \cdot\ (E,E) \to E$
    $\wedge, \vee\ (B,B) \to B$
    $\neg\ (B) \to B$
    $\varkappa(B,E,E) \to C$
end Σ

and the sorted system of variables

$$v\colon \{a, b, c, d, x, y, z\} \to \{E, C, B\}$$

with $av = bv = cv = dv = E$, $xv = yv = zv = B$.

Examples of Σ-terms on $v\colon X \to S$ are given by

$$tx \wedge \neg\, y \vee f \wedge \in T(\Sigma, v)_B,$$

$$ab1 + \cdot c + \in T(\Sigma, v)_E,$$

$$dd \cdot ca + \cdot \in T(\Sigma, v)_E,$$

$$xy \vee \neg\, ab + b \cdot abc \cdot \cdot 1 + \ [illegible]\ T(\Sigma, v)_C,$$

whereas $ax + \notin T(\Sigma, v)_E$, because xv [illegible] no[illegible] 1+ is also not a Σ-term on $v\colon X \to S$ be[illegible] , · is left out after the b or after the [illegible]

Now let A be the Σ-algebra [illegible] the set of natural numbers, and A_B [illegible] where

$t^A = T$, $f^A = F$, $0^A =$ zer[illegible]

$+^A =$ addition of na[illegible]

$\cdot^A =$ multiplicatio[illegible]

$\wedge^A =$ logical co[illegible]

$\vee^A =$ logical d[illegible]

$\neg^A =$ logica[illegible]

$\varkappa^A\colon \{T, F\}$ [illegible]

$(x, n_1, n$ [illegible]

Final[illegible] $h = 3$, $bh = 5$, $ch = 2$, $dh =$ [illegible] ermined homomorphism

co[illegible] evaluation of the formal ex- p[illegible] ion, with respect to the given a[illegible] ording to that

$$[illegible]\ \neg\, y \vee)\, h_B, fh_B\big) \wedge^A$$

$$[illegible]\ x \wedge \neg\, y \vee)\, h_B, F\big) \wedge^A = F,$$

$$[illegible]\ h_E = 3(5 + 1) + 2 = 20,$$

$$[illegible]\ a + \cdot)\, h_E = \big((0 \cdot 0) \cdot (2 + 3)\big) = 0,$$

$$(xy \vee \neg\, ab + b \cdot abc \cdot\cdot\, 1 + \varkappa)\, h_C$$

$$= \big((xy \vee \neg)\, h_B, (ab + b\, \cdot)\, h_E, (abc \cdot\cdot\, 1 +)\, h_E\big)\, \varkappa^A$$

$$= \big(\neg(T \vee F), (3 + 5) \cdot 5, 3 \cdot (5 \cdot 2) + 1\big)\, \varkappa^A = 3 \cdot (5 \cdot 2) + 1 = 31.$$

We see that the chosen notation of Σ-terms is not easy to read. If we want to assign to the Σ-terms a more readable notation, i.e. if we want to define another orthography, then we can do it by means of the recursion theorem. For that purpose we define the following *orthographical Σ-algebra* O, where

$$O_E = \{a, b, c, d, [,], +, 0, 1\}^+,$$

$$O_B = \{x, y, z, \wedge, \vee, \neg, [,], T, F\}^+,$$

$$O_C = \{a, b, c, d, x, y, z, +, \wedge, \vee, 0, 1, T, F, [,], \mathbf{IF}, \mathbf{THEN}, \mathbf{ELSE}, \neg\}^+$$

and where the Σ-indexed family of fundamental operations is given by the following string functions:

$$t^O = T, \quad f^O = F, \quad 0^O = 0, \quad 1^O = 1,$$

$$(w_1, w_2) +^O = [w_1 + w_2],$$

$$(w_1, w_2) \cdot^O = w_1 w_2,$$

$$(w_1, w_2) \vee^O = [w_1 \vee w_2],$$

$$(w_1, w_2) \wedge^O = [w_1 \wedge w_2],$$

$$w \neg^O = \neg [w],$$

$$(w_1, w_2, w_3)\, \varkappa^O = \mathbf{IF}\ w_1\ \mathbf{THEN}\ w_2\ \mathbf{ELSE}\ w_3 .$$

The last string function $\varkappa^O$ acts for instance as follows: write the character **IF**, append the first argument, append the character **THEN**, append the second argument, append the character **ELSE** and append finally the third argument. If we choose as assignment $\boldsymbol{i} \in O_v$ the inclusion, then the homomorphism

$$i: T(\Sigma, v) \to O$$

associates with the terms their new notation, so for instance

$$(xy \vee \neg\, ab + b \cdot abc \cdot \cdot 1 + \varkappa)\, i_C$$
$$= \mathbf{IF} \neg [x \vee y]\ \mathbf{THEN}\ [a + b]\, b\ \mathbf{ELSE}\ [abc + 1].$$

But, not every string of the carrier of O is an image with respect to the homomorphism $i: T(\Sigma, v) \to O$. In addition, one can see that different formal expressions can be mapped to the same notation, so for instance

$$(abc \cdot \cdot)\, i_E = (ab \cdot c \cdot)\, i_E .$$

By the preceding example we can illustrate one of the main problems of programing languages, namely the definition of syntax and semantics. With respect to our example we can say that the chosen orthography, represented by the Σ-algebra O, and the defined semantics, represented by the Σ-algebra A, are consistent, if $t_1 i_s = t_2 i_s$ for $s \in \{B, E, C\}$, $t_1, t_2 \in T(\Sigma, v)_s$ implies $t_1 h_s = t_2 h_s$ for every homomorphism

$$h: T(\Sigma, v) \to A$$

extending an assignment $\boldsymbol{h} \in A_v$, i.e. equal notation of formal expressions implies equal meaning. Precisely this principle was used in Section 1.3 for the definition of the semantics of the functional language.

Up to now we have only defined the meaning of a formal expression $t \in T(\Sigma, v)$ and not yet the meaning of an orthographical correct notation, i.e. of an element $w \in O_s$, $s \in \{B, C, E\}$, that is an image of a formal expression with respect to the homomorphism $i: T(\Sigma, v) \to O$. The answer to the last question can be given by the well-known theorem of homomorphisms. To formulate this theorem we need some further basic algebraic notions.

Definition 2.2.5. Let $\Sigma = (S, \alpha: \Omega \to S^* \times S)$ be any signature. A Σ-algebra A is called a *subalgebra* of a Σ-algebra B if

(1) $A_s \subseteq B_s$ for every $s \in S$, and

(2) the S-inclusion $i: A \to B$, i.e. $xi_s = x$ for every $s \in S$, $x \in A_s$, is a Σ-homomorphism. ∎

Obviously, for any S-subset $A \subseteq B$ of the underlying S-set of a Σ-algebra B there is at most one Σ-algebra with A as underlying S-set and which is a Σ-subalgebra of B. An S-subset $A \subseteq B$ of the S-carrier of a Σ-algebra B is an S-carrier of a Σ-subalgebra if and only if for every $\sigma \in \Omega$, where $\sigma: w \to s$, and for every $\boldsymbol{a} \in A_w$ the application of σ^B to $\boldsymbol{a}$ yields an element of A_s, i.e. A is closed with respect to the application of fundamental operation of B to arguments taken from A.

For the sake of completeness we summarize here some well-known facts of the theory of total many-sorted algebras without proof. Proofs and more details can be found in the book of Lugowski (1976) or in the paper of Birkhoff and Lipson (1970).

Definition 2.2.6. Let $A, B \in \mathrm{ALG}_t(\Sigma)$. For any Σ-homomorphism $f: A \to B$ the set-image $Af \subseteq B$ of the S-set A is an S-carrier of a Σ-subalgebra of B, which will be called the *homomorphic image of* A, denoted by Af. ∎

According to Definition 2.2.6 the $\{B, E, C\}$-set of orthographically correct representations of formal expressions is a carrier of a subalgebra of O.

Definition 2.2.7. For any $A \in \mathrm{ALG}_t(\Sigma)$ an S-indexed family $\varrho = (\varrho_s \mid s \in S)$ of equivalence relations $\varrho_s \subseteq A_s \times A_s$, $s \in S$, is called a *congruence relation* in A if $(\boldsymbol{x}\sigma^A, \boldsymbol{y}\sigma^A) \in \varrho_s$ holds for every $\sigma \in \Omega$, $\sigma: w \to s$, and for every $\boldsymbol{x}, \boldsymbol{y} \in A_w$ with $(i\boldsymbol{x}, i\boldsymbol{y}) \in \varrho_{iw}$ for every $i \in [n]$, where $w: [n] \to S$. ∎

Theorem 2.2.8. *For any Σ-homomorphism $f: A \to B$ the S-indexed family*

$$\ker f = (\ker f_s \mid s \in S)$$

with

$$\ker f_s = \{(x, y) \in A_s \times A_s \mid xf_s = yf_s\}, \qquad s \in S,$$

is a congruence relation in A. ∎

Theorem 2.2.9 (*Theorem of Homomorphisms*). *Let Σ-homomorphisms $f: B \to A$, $g: B \to C$ be given such that $g_s: B_s \to C_s$ is surjective for every $s \in S$, briefly $g: B \to C$ is surjective. Then there exists a Σ-homomorphism $h: C \to A$ with $f = g \circ h$ if and only if* $\ker g_s \subseteq \ker f_s$ *for every sort name $s \in S$.* ∎

The homomorphism theorem can be applied to the problem described above if we take $B = T(\Sigma, v)$, for A the semantic algebra, for C the homomorphic image $T(\Sigma, v)\, i$ of $i: T(\Sigma, v) \to 0$, i.e. the subalgebra of orthographically correct representations of formal expressions, and if we assume that $f: T(\Sigma, v) \to A$ is the extension of any assignment $\boldsymbol{f} \in A_v$, and that $g: T(\Sigma, v) \to T(\Sigma, v)\, i$ is the restriction of $i: T(\Sigma, v) \to 0$ to its homomorphic image. The condition that equal representations imply equal meaning then guarantees by Theorem 2.2.9 that a homomorphism $h: T(\Sigma, v)\, i \to A$ exists defining the meaning for every orthographically correct representation.

The logical structure of a compiler can be reconstructed from this algebraic definition of semantics by $h: T(\Sigma, v)\, i \to A$. The aim of syntactic analysis is to decide whether a given string is an orthographically correct representation of a formal expression and if so to construct the corresponding internal representation. Sometimes even the postfix Polish notation is used as the internal form of formal expressions. The second part of a compiler, the semantic interpretation, corresponds to a description of the semantic algebra in terms of the machine or assembler language of the target computer. The main point of this description is the definition of the fundamental operations of A by suitable programs.

Now that this example is complete we shall make a closer study of the relations between $T(\Sigma, v)$ and partial Σ-algebras for a general signature Σ and any S-sorted system of variables $v: X \to S$.

Definition 2.2.10. For any signature $\Sigma = (S, \alpha: \Omega \to S^* \times S)$ a *partial Σ-algebra*

$$A = \big((A_s \mid s \in S), (\sigma^A \mid \sigma \in \Omega)\big)$$

is given by an S-set A and by a Σ-indexed family of partial fundamental operations, so that the domain of the partial fundamental operation, σ^A, denoted by dom σ^A, is a subset of A_w if $\sigma: w \to s$, i.e.

$$\sigma^A: \operatorname{dom} \sigma^A \to A_s \quad \text{and} \quad \operatorname{dom} \sigma^A \subseteq A_w.$$

By ALG(Σ) we denote the class of all partial Σ-algebras. If $A, B \in \mathrm{ALG}(\Sigma)$, an S-mapping $f = (f_s: A_s \to B_s \mid s \in S)$ is called a *Σ-homomorphism* from A to B, denoted by $f: A \to B$, if for every $\sigma \in \Omega$, with $\sigma: w \to s$ and every $\boldsymbol{a} \in A_w$ the following conditions hold:

(1) $\boldsymbol{a} \in \operatorname{dom} \sigma^A$ implies $\boldsymbol{a} f_w \in \operatorname{dom} \sigma^B$.

(2) $\boldsymbol{a}\sigma^A f_s = \boldsymbol{a} f_w \sigma^B$ if both sides are defined.
A Σ-homomorphism $f: A \to B$ between partial Σ-algebras is called *closed* (strong in the sense of GRÄTZER (1978)) if additionally

(3) $\boldsymbol{a} f_w \in \operatorname{dom} \sigma^B$ implies $\boldsymbol{a} \in \operatorname{dom} \sigma^A$ for every $\boldsymbol{a} \in A_w$. ■

The various kinds of homomorphisms between partial algebras are extensively described by BURMEISTER (1986). We have introduced here only the weakest and the strongest versions of homomorphism, because they are the only ones we need for the concept of equational partiality. Since we are not interested in arbitrary partial algebras, we give no examples for Definition 2.2.10 here. Readers who are interested in this direction should consult the book of BURMEISTER.

In the second half of this section we introduce the concept of *existence equations for many-sorted algebras*. For the case of homogeneous algebras this concept is discussed by BURMEISTER in detail, so that we give here only a brief but complete introduction.

Definition 2.2.11. Let $v: X \to S$ be any S-sorted system of variables and $\Sigma = (S, \alpha: \Omega \to S^* \times S)$ any signature. If $\sigma \in \Omega$, $\sigma: w \to s$, $w: [n] \to S$, and

$$t = \boldsymbol{t}\sigma^{T(\Sigma, v)} \quad \text{for some} \quad \boldsymbol{t} \in T(\Sigma, v)_w$$

then every term $i\boldsymbol{t}$ for $i \in [n]$ is called an *immediate subterm* of $t \in T(\Sigma, v)_s$. We denote by $\leq$ the reflexive and transitive closure of the immediate subterm relation, and call it the *subterm relation*. ■

It is obvious that the subterm relation is a partial order on the union of the carriers of $T(\Sigma, v)$. The minimal elements of this partially ordered set are the variables and the constant operators. Additionally, the partially ordered set of all Σ-terms on $v: X \to S$ satisfies the decreasing chain condition, which means that any strictly decreasing chain

$$t_0 > t_1 > \ldots > t_n$$

has finite length.

Definition 2.2.12. Given $\Sigma = (S, \alpha: \Omega \to S^* \times S)$, $v: X \to S$, a partial Σ-algebra $A \in \mathrm{ALG}(\Sigma)$, and an assignment $\boldsymbol{a} \in A_v$, then we define inductively the S-set

$$\operatorname{dom} \tilde{\boldsymbol{a}} \subseteq T(\Sigma, v)$$

of all ***a**-interpretable terms*, and define also inductively the value

$$t\tilde{\boldsymbol{a}}_s \in A_s \quad \text{of the } \boldsymbol{a}\text{-interpretation of} \quad t \in T(\Sigma, v)_s$$

as follows:

(1) Every variable $x \in X$ is $\boldsymbol{a}$-interpretable, and $x\tilde{\boldsymbol{a}}_{xv} = x\boldsymbol{a}$ for every $x \in X$.

(2) Let $\sigma \in \Omega$, $\sigma: w \to s$, $w: [n] \to S$, $\boldsymbol{t} \in T(\Sigma, v)_w$ and $t = \boldsymbol{t}\sigma^{T(\Sigma,v)}$. Then t is $\boldsymbol{a}$-interpretable if and only if firstly $\boldsymbol{it}$ is $\boldsymbol{a}$-interpretable for every $i \in [n]$ and secondly $\boldsymbol{t}\tilde{\boldsymbol{a}}_w \in \operatorname{dom} \sigma^A$. The $\boldsymbol{a}$-interpretation of t is then defined by

$$(\boldsymbol{t}\sigma^{T(\Sigma,v)})\, \tilde{\boldsymbol{a}}_s = (\boldsymbol{t}\tilde{\boldsymbol{a}}_w)\, \sigma^A. \qquad \blacksquare$$

If $A \in \mathrm{ALG}_t(\Sigma)$, then any term is $\boldsymbol{a}$-interpretable and the unique extension of $\boldsymbol{a} \in A_v$ to a homomorphism coincides with the $\boldsymbol{a}$-interpretation

$$\tilde{\boldsymbol{a}}: \operatorname{dom} \tilde{\boldsymbol{a}} = T(\Sigma, v) \to A.$$

It is easy to see that the set of $\boldsymbol{a}$-interpretable terms is closed with respect to subterms. Furthermore, we mention that the homomorphism

$$\tilde{\boldsymbol{a}}: \operatorname{dom} \tilde{\boldsymbol{a}} \to A$$

is a closed homomorphism, and that the $\boldsymbol{a}$-interpretation is the greatest extension of $\boldsymbol{a} \in A_v$ to a closed homomorphism

$$\boldsymbol{a}^{\#}: B \to A,$$

where B is a partial Σ-algebra such that $B_s \subseteq T(\Sigma, v)_s$ for every $s \in S$, where the S-inclusion forms a homomorphism and the union of the s-carriers of B is closed with respect to subterms.

Definition 2.2.13. An ordered pair $(t_1, t_2) \in T(\Sigma, v)_s \times T(\Sigma, v)_s$, $s \in S$, of Σ-terms on $v: X \to S$ is said to be an *existence equation* on $v: X \to S$, denoted by

$$(v: t_1 \overset{\mathrm{e}}{=} t_2).$$

If A is any partial Σ-algebra, an assignment $\boldsymbol{a} \in A_v$ is called a *solution* of $(v: t_1 \overset{\mathrm{e}}{=} t_2)$ in A, if

$$t_1 \in (\operatorname{dom} \tilde{\boldsymbol{a}})_s, \quad t_2 \in (\operatorname{dom} \tilde{\boldsymbol{a}})_s \quad \text{and} \quad t_1\tilde{\boldsymbol{a}}_s = t_2\tilde{\boldsymbol{a}}_s.$$

This fact will be denoted by

$$(A, \boldsymbol{a}) \models (v: t_1 \overset{\mathrm{e}}{=} t_2).$$

We denote by $(v: G)$, where

$$G \subseteq \bigcup_{s \in S} T(\Sigma, v)_s \times T(\Sigma, v)_s,$$

a set of existence equations on $v: X \to S$.

For a set of existence equations $(v: G)$ and any partial Σ-algebra A we define the *set of solutions* of $(v: G)$ in A by

$$A_{(v:G)} = \{\boldsymbol{a} \in A_v \mid (A, \boldsymbol{a}) \vdash (v: t_1 \overset{\mathrm{e}}{=} t_2) \quad \text{for all} \quad (v: t_1 \overset{\mathrm{e}}{=} t_2) \in G\},$$
$$A_{(v:\emptyset)} = A_v.$$

It is useful to mention here that the concept of existence equations is not powerful enough to describe the interesting laws interrelating partial operations. This is not caused by a wrong concept of validity of equations for partial algebras. Up to now no concept of validity of equations is known which is capable of defining the class of all small categories as a class of equationally definable partial algebras.

We illustrate the concept of $\boldsymbol{a}$-interpretation in partial algebras on the basis of the following partial algebra M for the example of a signature defined above:

$M_B = \{T, F\}$ the set of truth values;

$M_E = M_C =$ set of all (n,k)-matrices, where $n, k \in \mathbb{N}$ are not fixed, with natural numbers as elements;

$t^M = T,\ f^M = F,\ 0^M = (0)_{1,1},\ 1^M = \begin{pmatrix} 1 & 0 \\ 0 & 1 \end{pmatrix}_{2,2}.$

$\wedge^M$ = logical conjunction;

$\vee^M$ = logical disjunction;

$\neg^M$ = logical negation;

$+^M$ = addition of matrices;

$\circ^M$ = multiplication of matrices;

$$(x, m_1, m_2)\,\varkappa^M = \begin{cases} m_1 & \text{if } x = T \\ m_2 & \text{if } x = F \end{cases} \quad \text{for all} \quad m_1, m_2 \in M_E = M_C.$$

An assignment $\boldsymbol{h} \in M_v$ may be given by $x\boldsymbol{h} = y\boldsymbol{h} = T$, $z\boldsymbol{h} = F$, $a\boldsymbol{h}$ = any (1,2)-matrix, $b\boldsymbol{h}$ = any (3,4)-matrix, $c\boldsymbol{h}$ = any (5,5)-matrix, and $d\boldsymbol{h}$ = any (2,3)-matrix.

Now the binary operations $+^M$, $\circ^M$ are partial ones, because an (n_1,k_1)-matrix m_1 can be added to an (n_2, k_2)-matrix m_2 iff $n_1 = n_2$ and $k_1 = k_2$, and the product $(m_1, m_2) \circ^M$ exists iff $n_2 = k_1$. It is easy to see that the following formal expressions are $\boldsymbol{h}$-interpretable in M: $11\circ$, $11\circ1\circ d\circ$, $db\circ$, $db\circ db\circ+$, $ccc++c\circ$, whereas the following formal expressions are not $\boldsymbol{h}$-interpretable in M: $01+$, $bc+$, $bc\circ$, $bd\circ$.

Another notion, important for technical reasons, is the *simultaneous substitution of variables* in a formal expression by given formal expressions. To be more precise, let be given S-sorted systems of variables $x: X \to S$, $y: Y \to S$, $z: Z \to S$ and term assignments $\boldsymbol{t}_1 \in T(\Sigma, y)_x$, $\boldsymbol{t}_2 \in T(\Sigma, z)y$. We make the situation clearer by the following diagram:

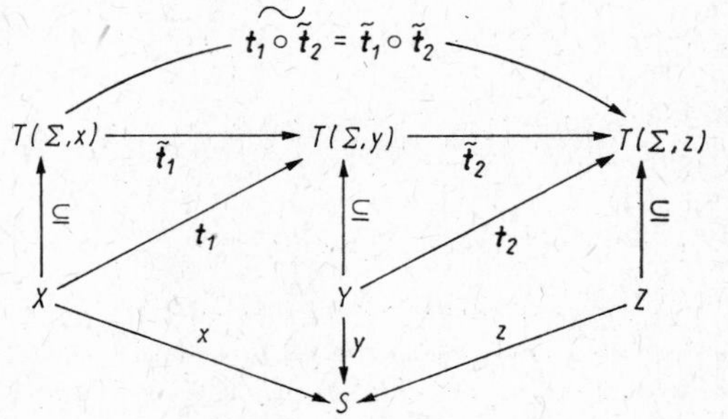

Since $T(\Sigma, y)$, $T(\Sigma, z)$ are total Σ-algebras dom $\boldsymbol{t}_1 = T(\Sigma, x)$ and dom $\boldsymbol{t}_2 = T(\Sigma, y)$. The assignment

$$\boldsymbol{t}_1 \circ \tilde{\boldsymbol{t}}_2 \in T(\Sigma, z)_x$$

results from $\boldsymbol{t}_1$ by simultaneous substitution of the variables contained in $x\boldsymbol{t}_1 \in T(\Sigma, y)$, $x \in X$, according to the assignment $\boldsymbol{t}_2 \in T(\Sigma, z)_y$. Additionally holds

$$\widetilde{\boldsymbol{t}_1 \circ \tilde{\boldsymbol{t}}_2} = \tilde{\boldsymbol{t}}_1 \circ \tilde{\boldsymbol{t}}_2 .$$

We extend this notion from terms and term assignments to existence equations and sets of existence equations. Given an existence equation $(v\colon t_1 \overset{e}{=} t_2)$ or a set of existence equations $(v\colon G)$, and an assignment $\boldsymbol{t} \in T(\Sigma, w)_v$, then

$$(v\colon t_1 \overset{e}{=} t_2)\, \tilde{\boldsymbol{t}} = (w\colon t_1\tilde{\boldsymbol{t}} \overset{e}{=} t_2\tilde{\boldsymbol{t}}),$$

$$(v\colon \emptyset)\, \tilde{\boldsymbol{t}} = (w\colon \emptyset),$$

$$\begin{aligned}(v\colon G)\, \tilde{\boldsymbol{t}} &= \{(v\colon t_1 \overset{e}{=} t_2)\, \tilde{\boldsymbol{t}} \mid (v\colon t_1 \overset{e}{=} t_2) \in G\}\\ &= \{(w\colon t_1\tilde{\boldsymbol{t}} \overset{e}{=} t_2\tilde{\boldsymbol{t}}) \mid (v\colon t_1 \overset{e}{=} t_2) \in G\}.\end{aligned}$$

We conclude this section by a further technical notion. For any term $t \in T(\Sigma, v)$, $v\colon X \to S$, and any set of existence equations $(v\colon G)$ we define the sets of operators

$$\mathrm{op}(t) \subseteq \Omega \quad \text{and} \quad \mathrm{op}(v\colon G) \subseteq \Omega$$

as the sets of operators syntactically contained in t and in $(v\colon G)$ respectively:

$$\mathrm{op}(x) = \emptyset \quad \text{for every} \quad x \in X,$$

$$\mathrm{op}(\boldsymbol{t}) = \bigcup_{i \in [n]} \mathrm{op}(i\boldsymbol{t}) \quad \text{for} \quad \boldsymbol{t} \in T(\Sigma, v)_w, \quad w\colon [n] \to S,$$

$$\mathrm{op}(t) = \mathrm{op}(\boldsymbol{t}) \cup \{\sigma\}, \quad \text{for} \quad t = \boldsymbol{t}\sigma^{T(\Sigma,v)} \quad \text{with} \quad \sigma \in \Omega,$$

$$\sigma\colon w \to s, \qquad \boldsymbol{t} \in T(\Sigma, v)_w,$$

$$\mathrm{op}(\boldsymbol{t}) = \bigcup_{y \in Y} \mathrm{op}(y\boldsymbol{t}) \quad \text{for} \quad \boldsymbol{t} \in T(\Sigma, v)_w, \quad w\colon Y \to S,$$

$$\mathrm{op}(v\colon t_1 \overset{e}{=} t_2) = \mathrm{op}(t_1) \cup \mathrm{op}(t_2),$$

$$\mathrm{op}(v\colon G) = \bigcup \{\mathrm{op}(v\colon t_1 \overset{e}{=} t_2) \mid (v\colon t_1 \overset{e}{=} t_2) \in G\}.$$

2.3. Equationally partial heterogeneous algebras

As mentioned above, we describe the partiality by so-called *domain conditions,* associated with every operator which are necessary and sufficient for the applicability of the operator. Recall the example M of Section 2.2

where $+^M$ is the addition of matrices and $\circ^M$ is the multiplication of matrices with natural numbers as elements. It is well-known that both $+^M$ and $\circ^M$ have a necessary and sufficient domain condition:

$(m_1, m_2) +^M$ is defined iff the number of rows and columns of the matrices m_1 and m_2 are equal, and

$(m_1, m_2) \circ^M$ is defined iff the number of columns of m_1 equals the number of rows of m_2.

But these domain conditions are formulated by means of natural numbers. Does this mean that the concept of natural numbers has to be included into the theory of partial algebras? One should be very careful when enriching the theory of partial algebras with any additional tools. The theory should be technically as simple as possible. One can expect that other examples would require other additional concepts if one wants to formulate the usual domain conditions directly.

Therefore we look for another formulation of the domain condition of $+^M$ and $\circ^M$. The concept of natural numbers can be reformulated in terms of matrices themselves using the unit matrices $(1)_{n,n}$. So, we can define additional unary operations

$$l^M: M_E \to M_E, \qquad r^M: M_E \to M_E$$

such that for any (n,k)-matrix $m \in M_E$

$$ml^M = (n,n)\text{-unit-matrix},$$

$$mr^M = (k,k)\text{-unit-matrix}.$$

Using these operations we can reformulate the domain conditions as follows

$$(m_1, m_2) +^M \quad \text{is defined iff} \quad m_1 l^M = m_2 l^M, \quad \text{and} \quad m_1 r^M = m_2 r^M,$$

$$(m_1, m_2) \circ^M \quad \text{is defined iff} \quad m_1 r^M = m_2 l^M.$$

We pay attention to the fact that now term equations serve as domain conditions.

Readers familiar with category theory will not be surprised by this fact, since the algebra M is derived from some category of linear mappings.

For small categories the domain condition of the partial binary multiplication (i.e. composition of morphisms) is a term equation, too.

Appealing to these examples, we adopt the convention that domain conditions are always sets of existence equations.

Definition 2.3.1. An *equationally partial signature*, briefly *ep-signature*

$$\theta = \big(S, \alpha\colon \Omega \to S^* \times S, (\mathrm{def}\,\sigma \mid \sigma \in \Omega)\big)$$

is given by an operator scheme, i.e. by a signature $(S, \alpha\colon \Omega \to S^* \times S)$, and a family $(\mathrm{def}\,\sigma \mid \sigma \in \Omega)$ of sets of existence equations, so that for every $\sigma \in \Omega$ where $\sigma\colon w \to s$, the domain condition def σ is a set of existence equations on $w \in S^*$. A partial $(S, \alpha\colon \Omega \to S^* \times S)$-algebra

$$A = \big((A_s \mid s \in S), (\sigma^A \mid \sigma \in \Omega)\big)$$

is said to be an *equationally partial many-sorted algebra* of type θ, briefly *θ-algebra*, if

$$\mathrm{dom}\,\sigma^A = A_{\mathrm{def}\,\sigma}$$

for every $\sigma \in \Omega$. ■

Sometimes

$$\theta = \big(\Sigma, (\mathrm{def}\,\sigma \mid \sigma \in \Omega)\big)$$

denotes an equationally partial signature, where

$$\Sigma = (S, \alpha\colon \Omega \to S^* \times S)$$

denotes the signature associated with θ. $\mathrm{ALG}\big(\Sigma, (\mathrm{def}\,\sigma \mid \sigma \in \Omega)\big)$ denotes the class of all θ-algebras just as $\mathrm{ALG}(\theta)$ denotes that class.

It is very useful to extend the notion of domain conditions from operators to terms. For that reason we extend the notion $\sigma\colon w \to s$ from operators to terms, too. If $v\colon X \to S$ and $w\colon Y \to S$ are S-sorted systems of variables, then

$$t\colon w \to s \quad \text{abbreviates} \quad t \in T(\Sigma, w)_s,$$

and

$$\boldsymbol{t}\colon w \to v \quad \text{abbreviates} \quad \boldsymbol{t} \in T(\Sigma, w)_v.$$

Definition 2.3.2. *Domain conditions of terms* can be defined by induction as follows:

(a) def $x = (v{:}\emptyset)$ for every $x \in X$, where $v\colon X \to S$ and $x\colon v \to xv$ is considered as a term;

(b) If $\sigma \in \Omega$, $\sigma\colon \lambda \to s$ then

$$\mathrm{def}\,\sigma^{T(\Sigma,v)} = (\mathrm{def}\,\sigma)\,\tilde{\boldsymbol{t}}$$

where $\boldsymbol{t}\colon v \to \lambda$ is the only assignment of the empty string in $T(\Sigma, v)$;

(c) If $\sigma \in \Omega$, $\sigma\colon w \to s$, $w\colon [n] \to S$, $n \neq 0$, $\boldsymbol{t}\colon v \to w$ and $t = \boldsymbol{t}\sigma^{T(\Sigma,v)}$, then

$$\mathrm{def}\,t = (\mathrm{def}\,\sigma)\,\tilde{\boldsymbol{t}} \cup \bigcup_{i\in[n]} \mathrm{def}\,i\boldsymbol{t}.$$

For any existence equation $(v\colon t_1 \overset{e}{=} t_2)$ we define its domain condition $\operatorname{def}(v\colon t_1 \overset{e}{=} t_2)$ as the union of the domain conditions of t_1 and t_2. For any $(v{:}G)$ we define its domain condition $\operatorname{def}(v{:}G)$ as the union of the domain conditions of its elements. Finally we set $\operatorname{def}(v{:}\emptyset) = (v{:}\emptyset)$. ■

It is not only useful to extend notions concerning operators to terms but most properties of operations can be generalized inductively to terms. We underline this remark by the following corollary. To formulate this corollary we have to extend the interpretation of operators by their defining operations to terms by so-called *term functions*. Thus, we associate with every term $t\colon v \to s$, where $v\colon X \to S$ and $s \in S$, and every partial Σ-algebra A a term function

$$t^A\colon \operatorname{dom} t^A \to A_s \quad \text{with} \quad \operatorname{dom} t^A \subseteq A_v.$$

Definition 2.3.3. Let be given $\theta = \big(\Sigma, (\operatorname{def} \sigma \mid \sigma \in \Omega)\big)$, $v\colon X \to S$ and $A \in \mathrm{ALG}(\Sigma)$. Then we define:

(a) With any $x \in T(\Sigma, v)_{xv}$ we associate

$$x^A\colon A_v \to A_{xv}$$

so that $\boldsymbol{a}(x^A) = x\boldsymbol{a}$ for every $\boldsymbol{a} \in A_v$.

(b) $\sigma\colon \lambda \to s$. Although in this case the fundamental operation σ^A and the associated terms function are different mappings, we will use the same notation. It is always clear from context, whether σ^A denotes the fundamental operation of A or the term function. The mappings are different only in their domains. Now we define the terms function $\sigma^A\colon \operatorname{dom} \sigma^A \to A_s$. We set $\operatorname{dom} \sigma^A = \emptyset$ if the value of the fundamental operation σ^A is undefined, and if σ^A is defined in A we set $\operatorname{dom} \sigma^A = A_v = \{\emptyset\}$ and $\boldsymbol{a}\sigma^A = \sigma^A$.

(c) Let be $\sigma \in \Omega$, $\sigma\colon w \to s$, $w\colon [n] \to S$, $n \neq 0$, $\boldsymbol{t} \in T(\Sigma, v)_w$ and $t = \boldsymbol{t}\sigma^{T(\Sigma,v)}$. Then we set

$$\operatorname{dom} t^A = \big\{\boldsymbol{a} \in A_v \mid \boldsymbol{a} \in \operatorname{dom}(i\boldsymbol{t})^A \quad \text{for every} \quad i \in [n]$$
$$\text{and} \quad \big(\boldsymbol{a}(1\boldsymbol{t})^A, \ldots, \boldsymbol{a}(n\boldsymbol{t})^A\big) \in \operatorname{dom} \sigma^A\big\}$$

and

$$\boldsymbol{a}t^A = \big(\boldsymbol{a}(1\boldsymbol{t})^A, \ldots, \boldsymbol{a}(n\boldsymbol{t})^A\big)\, \sigma^A$$ ■

By a very simple induction one can prove

Corollary 2.3.4.

(a) $\operatorname{dom} t^A = \{\boldsymbol{a} \in A_v \mid t \in \operatorname{dom} \tilde{\boldsymbol{a}}\}$
for every $t \in T(\Sigma, v)_s$, $v\colon X \to S$, *and* $s \in S$.

(b) $\operatorname{dom} \tilde{\boldsymbol{a}} \subseteq \operatorname{dom} \widetilde{\boldsymbol{a}f_v}$ *and* $(t\tilde{\boldsymbol{a}}_s)\, f_s = t(\widetilde{\boldsymbol{a}f_v})_s$
for every $\boldsymbol{t} \in T(\Sigma, v)_s$, $v\colon X \to S$, $s \in S$, $\boldsymbol{a} \in \operatorname{dom} t^A$ *and every homomorphism* $f\colon A \to B$. ■

The conditions of (b) can be visualized by the following diagram:

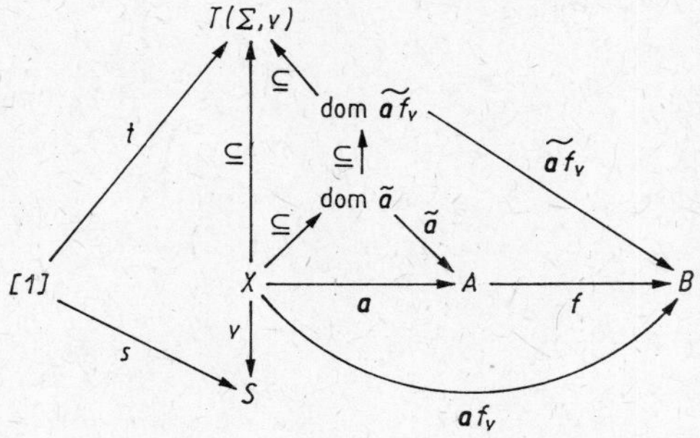

The meaning of a term function t^A can more easily be derived from Definition 2.3.3, but for dealing with term functions in proofs the representation in Corollary 2.3.4 is of great value.

By the following corollary we state some properties of term functions.

Corollary 2.3.5.

(a) *A partial Σ-algebra* $A \in \mathrm{ALG}(\Sigma)$ *is an equationally partial algebra* $A \in \mathrm{ALG}\big(\Sigma, (\mathrm{def}\,\sigma \mid \sigma \in \Omega)\big)$, *if and only if*

$$\mathrm{dom}\, t^A = A_{\mathrm{def}t}$$

for every term $t \in T(\Sigma, v)$, $v\colon X \to S$.

(b) *If* $A, B \in \mathrm{ALG}(\Sigma)$, $f = (f_s\colon A_s \to B_s \mid s \in S)$, *then f is a homomorphism* $f\colon A \to B$, *if and only if for every* $v\colon X \to S$, $t \in T(\Sigma, v)$, $\boldsymbol{a} \in A_v$ *it holds that* $\boldsymbol{a} \in \mathrm{dom}\, t^A$ *implies* $\boldsymbol{a} f_v \in \mathrm{dom}\, t^B$ *and*

$$(\boldsymbol{a} t^A)\, f_s = (\boldsymbol{a} f_v)\, t^B.$$

Proof. At first we assume $A \in \mathrm{ALG}\big(\Sigma, (\mathrm{def}\,\sigma \mid \sigma \in \Omega)\big)$ and show $\mathrm{dom}\, t^A = A_{\mathrm{def}t}$ inductively. Therefore, let $t = x$. Then $\mathrm{dom}\, t^A = A_v$ holds according to (a) in Definition 2.3.3, and because $\mathrm{def}\, x = (v{:}\emptyset)$ we obtain $\mathrm{dom}\, t^A = A_v = A_{(v:\emptyset)} = A_{\mathrm{def}t}$. Now let $t = \boldsymbol{t}\sigma^{T(\Sigma,v)}$ and $\mathrm{dom}\,(i\boldsymbol{t})^A = A_{\mathrm{def}i\boldsymbol{t}}$ for every $i \in [n]$. If $\boldsymbol{a} \in \mathrm{dom}\, t^A$ then we can infer as follows according to (b) and (c) in Definition 2.3.3 and by means of Section 2.2:

$$\begin{aligned}
\mathrm{dom}\, t^A &= \bigcap_{i \in [n]} \mathrm{dom}(i\boldsymbol{t})^A \cap \{\boldsymbol{a} \in A_v \mid \boldsymbol{t}\boldsymbol{a}_w \in \mathrm{dom}\, \sigma^A\} \\
&= \bigcap_{i \in [n]} A_{\mathrm{def}i\boldsymbol{t}} \cap \{\boldsymbol{a} \in A_v \mid \boldsymbol{t}\boldsymbol{a}_w \in A_{\mathrm{def}\sigma}\} \\
&= \bigcap_{i \in [n]} A_{\mathrm{def}i\boldsymbol{t}} \cap \{\boldsymbol{a} \in A_v \mid \boldsymbol{a} \in A_{(\mathrm{def}\sigma)\boldsymbol{t}}\} \\
&= A_{\left(\bigcup_{i \in [n]} \mathrm{def}i\boldsymbol{t} \cup (\mathrm{def}\sigma)\boldsymbol{t}\right)} = A_{\mathrm{def}t}.
\end{aligned}$$

The converse direction is evident since every operator $\sigma \in \Omega$ with $\sigma: w \to s$, $w: [n] \to S$ can be identified with the term

$$\boldsymbol{x}\sigma^{T(\Sigma,v)},$$

where $i\boldsymbol{x} = i$ for every $i \in [n]$. With the postfix Polish notation of terms we get

$$\boldsymbol{x}\sigma^{T(\Sigma,v)} = 12 \ldots n\sigma \quad \text{and} \quad (12 \ldots n\sigma)^A = \sigma^A.$$

Secondly we assume that $f: A \to B$ is a homomorphism. Because of Corollary 2.3.4 $\boldsymbol{a} \in \operatorname{dom} t^A$ yields $t \in \operatorname{dom} \tilde{\boldsymbol{a}} \subseteq \operatorname{dom} \widetilde{\boldsymbol{a}f_v}$, so that $\boldsymbol{a}f_v \in \operatorname{dom} t^B$, and

$$(\boldsymbol{a}t^A)\, f_s = (t\tilde{\boldsymbol{a}}_s)\, f_s = t(\tilde{\boldsymbol{a}}_s f_s) = t(\widetilde{\boldsymbol{a}f_v})_s = (\boldsymbol{a}f_v)\, t^B. \qquad \blacksquare$$

Now we are in a position to continue the concept of equational partiality. For this reason we consider the following ep-signature

$$\theta_0 = \big(\Sigma, (\operatorname{def} \sigma \mid \sigma \in \Omega)\big)$$

such that $\operatorname{def} \sigma = (w: \{12 \ldots n\sigma \stackrel{e}{=} 12 \ldots n\sigma\})$ for every $\sigma \in \Omega$, where $\sigma: w \to s$, $w: [n] \to S$.

Clearly, every partial Σ-algebra is a θ_0-algebra, too. If $A \in \mathrm{ALG}(\Sigma)$, $\sigma \in \Omega$ with $\sigma: w \to s$ then

$$A_{(w:\{12\ldots n\sigma \stackrel{e}{=} 12\ldots n\sigma\})} = \{\boldsymbol{a} \in A_w \mid \boldsymbol{a} \in \operatorname{dom} \sigma^A\}$$

so this kind of domain condition implies no restriction whatever.

This example shows that the concept of equational partiality as developed up to this point does not exclude the problems and difficulties of unrestricted partial algebras investigated in Burmeister (1986).

By the next example we show that the concept of a θ-subalgebra cannot be reduced to the requirement that the inclusion is a homomorphism. We will show that on one and the same S-subset non-isomorphic θ-subalgebras exist.

In connection with this example we demonstrate our convention for representing equationally partial operators.

θ_1 **is sorts** M
  oprn $k \to M$
    $f(x{:}\ M$ **iff** $xg{=}k) \to M$
    $g(x{:}M$ **iff** $xf{=}k) \to M$
end θ_1

we consider the θ_1-algebras A, B with

$$A_M = B_M = \{a\},$$
$$k^A = k^B = a, \qquad f^B = g^B = \mathrm{Id}_{\{a\}},$$
$$\operatorname{dom} f^A = \operatorname{dom} g^A = \emptyset.$$

It is evident that the total algebra B is really a θ_1-algebra. The set of solutions in A of the domain condition

$$\text{def}\, f = (w\colon \{x\} \to \{M\}\colon \{xg \overset{\text{e}}{=} k\})$$

consists of all those elements $a \in A_M$ such that ag^A exists and equals $a = k^A \in A_M$. Due to dom $g^A = \emptyset$ the set of solutions of def f in A is empty, too. In the same way follows $A_{\text{def}g} = \emptyset$. Therefore A is also a θ_1-algebra. Both A and B are subalgebras of the θ_1-algebra C given by

$$C_M = \{a, b\}, \qquad k^C = a, \qquad f^C = g^C\colon \{a, b\} \to \{a, b\}$$

with

$$xf^C = xg^C = a \quad \text{for each} \quad x \in C_M\,.$$

In the terminology of the general theory of partial algebras, investigated in BURMEISTER (1986), B is a subalgebra of C and A is a relative subalgebra of C.

The situation is quite different if we deal with the ep-signature θ_{CAT} corresponding to small categories:

> θ_{CAT} **is sorts** Obj, Mor
>   **oprn** id(Obj) → Mor
>   source(Mor) → Obj
>   target(Mor) → Obj
>   comp(x:Mor,y:Mor **iff** x target = y source) → Mor
> **end** θ_{CAT}

In this case one can prove that any injective homomorphism $f\colon A \to B$ between θ_{CAT}-algebras reflects solutions of the domain condition def comp, so that any injective homomorphism is a closed homomorphism (see Definition 2.2.10). Additionally, this implies that any bijective homomorphism is an isomorphism.

The examples above make it clear that we have to enrich the concept of an ep-signature if we want to exclude all the problems of unrestricted partial algebras.

If we look for a property of the ep-signature θ_{CAT} which makes it qualitatively different from θ_0 and θ_1, we notice that in the ep-signatures θ_0 and θ_1 the operators are immediately or indirectly used in their own domain conditions, whereas in θ_{CAT} some kind of hierarchy is given.

The aim of the following section is the investigation of this hierarchy condition.

2.4. Hierarchy conditions

First of all we will give the concept of a hierarchical ep-signature a precise meaning. It must be kept clear in the sequel what is meant by the statement that no operator is used immediately or indirectly in its own domain condition. For this reason we use the set op t, op(v:G) of operators which are syntactically used in any term $t \in T(\Sigma, v)$ and in any set of existence equations respectively, as defined at the end of Section 2.2.

Applying these definitions to the ep-signature θ_0, θ_1, and θ_{CAT} one gets

$$\text{op}(w\colon \{12 \ldots n\sigma = 12 \ldots n\sigma\}) = \{\sigma\}$$
$$\text{op}(\text{def } f) = \text{op}(w\colon \{x\} \to \{M\}\colon \{xg = k\}) = \{g, k\}$$
$$\text{op}(\text{def comp}) = \{\text{source, target}\}$$
$$\text{op}(\text{def id}) = \emptyset .$$

Definition 2.4.1. An ep-signature $\theta = (\Sigma, (\text{def } \sigma \mid \sigma \in \Omega))$ is said to be *hierarchical* if every sequence of not necessarily pairwise distinct operators

$$\sigma_0, \sigma_1, \sigma_2, \ldots$$

with $\sigma_{i+1} \in \text{op}(\text{def } \sigma_1)$ for every $i \in \mathbb{N}$ terminates, or in other words, there e a natural number n with $\text{op}(\text{def } \sigma_n) = \emptyset$.

Hierarchical ep-signature will be called *hep-signatures* for short. ■

Clearly, for the ep-signature θ_0 we can build an infinite sequence for every operator $\sigma \in \Omega_{\theta_0}$, namely $\sigma, \sigma, \sigma, \ldots$ because $\sigma \in \text{op}(\text{def } \sigma)$, so that θ_0 is not a hierarchical ep-signature.

For the ep-signature θ_1 we can construct the infinite sequence

$$f, g, f, g, \ldots, f, g, \ldots$$

because $f \in \text{op}(\text{def } g)$ and $g \in \text{op}(\text{def } f)$, so that θ_1 is also not hierarchical.

As mentioned above θ_{CAT} is hierarchical since any appropriate sequence of operators has at most the length two for op(def id) = op(def source) = op(def target) = $\emptyset$ and op(def comp) = {source, target}.

The importance of the concept of hep-signatures becomes evident by virtue of the following theorem.

Theorem 2.4.2. *For any ep-signature* $\theta = (\Sigma, (\text{def } \sigma \mid \sigma \in \Omega))$ *and θ-algebra A, B the following conditions are equivalent:*

(i) *θ is hierarchical;*

(ii) *Every bijective homomorphism $f\colon A \to B$ is an isomorphism;*

(iii) *Every injective homomorphism $f\colon A \to B$ reflects solutions of existenc equations;*

(iv) *Every injective homomorphism $f\colon A \to B$ reflects solutions of domain conditions.*

Before proving Theorem 2.4.2 we will make more precise the notion that a homomorphism $f\colon A \to B$ reflects solutions of existence equations and of sets of existence equations respectively.

Let $(v: t_1 \stackrel{e}{=} t_2)$ and $(v:G)$ be any existence equation and any set of existence equations respectively. A homomorphism $f: A \to B$ reflects solutions of $(v: t_1 \stackrel{e}{=} t_2)$ or of $(v:G)$ if for every assignment $\boldsymbol{a} \in A_v$

$$\boldsymbol{a}f_v \in B_{(v:t_1 \stackrel{e}{=} t_2)} \quad \text{implies} \quad \boldsymbol{a} \in A_{(v:t_1 \stackrel{e}{=} t_2)}$$

or

$$\boldsymbol{a}f_v \in B_{(v:G)} \quad \text{implies} \quad \boldsymbol{a} \in A_{(v:G)}.$$

Proof of Theorem 2.4.2: First we prove that (iv) implies (ii). Therefore let $f: A \to B$ be a bijective homomorphism. Then there exists the inverse S-mapping

$$f^{-1} = (f_s^{-1}: B_s \to A_s \mid s \in S).$$

We have to show that f^{-1} is a homomorphism. Let be $\sigma \in \Omega$, $\sigma: w \to s$, $\boldsymbol{b} \in B_{\mathrm{def}\sigma}$ and $\boldsymbol{a} = \boldsymbol{b}f_w^{-1} \in A_w$. Because of $\boldsymbol{a}f_w = \boldsymbol{b} \in B_{\mathrm{def}\sigma}$ it follows according to (iv) that $\boldsymbol{a} = \boldsymbol{b}f_w^{-1} \in A_{\mathrm{def}\sigma}$. The injectivity of the homomorphism $f: A \to B$ yields

$$\boldsymbol{b}\sigma^B f_s^{-1} = \boldsymbol{b}f_w^{-1}\sigma^A$$

because

$$\boldsymbol{b}\sigma^B f_s^{-1} f_s = \boldsymbol{b}\sigma^B = \boldsymbol{a}f_w\sigma^B = \boldsymbol{a}\sigma^A f_s = \boldsymbol{b}f_w^{-1}\sigma^A f_s.$$

This proves that $f^{-1}: B \to A$ is a homomorphism.

Second we show indirectly that (i) is a consequence of (ii). For that purpose we define $\Omega_k \subseteq \Omega$ as the set of all those operators $\sigma \in \Omega$ for which there is a non-terminating sequence

$$\sigma = \sigma_0, \sigma_1, \ldots, \sigma_n, \ldots$$

with $\sigma_{1+i} \in \mathrm{op}(\mathrm{def}\, \sigma_i)$ for $i = 0, 1, 2, \ldots$

If we assume that θ is not hierarchical, then $\Omega_k \neq \emptyset$. Analogously to the θ_1-algebras in Section 2.3 we define two θ-algebras F and F^* with $F_s^* = F_s = \{\emptyset\}$ for every $s \in S$. F is the uniquely determined total θ-algebra, e.g.

$$\mathrm{dom}\, \sigma^F = F_w \quad \text{for every} \quad \sigma \in \Omega \quad \text{with} \quad \sigma: w \to s.$$

The fundamental operations of F^* are given by

$$\mathrm{dom}\, \sigma^{F^*} = \begin{cases} F_w^* & \text{for every} \quad \sigma \in \Omega \setminus \Omega_k, \quad \text{where} \quad \sigma: w \to s \\ \emptyset & \text{for every} \quad \sigma \in \Omega_k. \end{cases}$$

The S-function

$$e = (e_s: F^* \to F_s \mid s \in S)$$

with $e_s = \mathrm{Id}_{\{\emptyset\}}$ for every $s \in S$ is a bijective homomorphism $e: F^* \to F$, however it is not an isomorphism. This contradiction to (ii) can only be resolved if $\Omega_k = \emptyset$, which means that θ is hierarchical.

It is not hard to show that (iv) holds provided (iii) is true. If def $\sigma = (w:\emptyset)$, where $\sigma: w \to s$, then $B_{\mathrm{def}\sigma} = B_w$ for every θ-algebra, so that for $\boldsymbol{a} \in A_w$ with $\boldsymbol{a}f_w \in B_{\mathrm{def}\sigma}$ obviously $\boldsymbol{a} \in A_{\mathrm{def}\sigma} = A_w$. Now we assume def $\boldsymbol{a} \neq (w: \emptyset)$, $\boldsymbol{a} \in A_w$ and $\boldsymbol{a}f_w \in B_{(w:t_1 \overset{e}{=} t_2)}$ for every $(w: t_1 \overset{e}{=} t_2) \in \mathrm{def}\,\sigma$, so that by (iii) $\boldsymbol{a} \in A_{(w:t_1 = t_2)}$ for every $(w: t_1 \overset{e}{=} t_2) \in \mathrm{def}\,\sigma$. This means $\boldsymbol{a} \in A_{\mathrm{def}\sigma}$ which is what we have to show.

Next we turn to the implication (iv) $\to$ (iii). By a simple induction one can prove that (iv) implies the following slightly generalized condition

(iv)* Every injective homomorphism $f: A \to B$ reflects solutions of def t for any term $t \in T(\Sigma, u)_s$, $u: Y \to S$.

The proof goes like the proof of Corollary 2.3.5. Let $(v: t_1 \overset{e}{=} t_2)$ with $v: X \to S$ be any existence equation, $\boldsymbol{a} \in A_v$ and $f: A \to B$ any injective homomorphism with

$$\boldsymbol{a}f_v \in B_{(v:t_1 \overset{e}{=} t_2)}.$$

This implies $\boldsymbol{a}f_v \in B_{\mathrm{def}t_1}$, $\boldsymbol{a}f_v \in B_{\mathrm{def}t_2}$, and $\boldsymbol{a}f_v t_1^B = \boldsymbol{a}f_v t_2^B$. Condition (iv)* together with the definition of the term functions t_1^B, t_2^B yield

$$\boldsymbol{a} \in A_{\mathrm{def}t_1}, \quad \boldsymbol{a} \in A_{\mathrm{def}t_2}, \quad \text{and} \quad t_1(\widetilde{\boldsymbol{a}f_v}) = t_2(\widetilde{\boldsymbol{a}f_v}).$$

Since $f: A \to B$ is a homomorphism we obtain by (2.3.4)

$$t_1\tilde{\boldsymbol{a}}_s f_s = t_1(\widetilde{\boldsymbol{a}f_v}) = t_2(\widetilde{\boldsymbol{a}f_v}) = t_2\tilde{\boldsymbol{a}}_s f_s,$$

where $s \in S$ is the sort name with $t_1, t_2 \in T(\Sigma, v)_s$.
The injectivity of $f_s: A_s \to B_s$ yields

$$t_1\tilde{\boldsymbol{a}} = t_2\tilde{\boldsymbol{a}}, \quad \text{so that} \quad \boldsymbol{a}t_1^A = \boldsymbol{a}t_2^A$$

or in other words

$$\boldsymbol{a} \in A_{(v:t_1 \overset{e}{=} t_2)}.$$

We conclude the proof of Theorem 2.4.2 by showing that (i) implies (iv). This is the main part of the proof. The presupposed hierarchy of θ allows us to prove (iv) by the following principle of induction: Let Ω be a poset (partially ordered set) in which every descending chain is finite. Let P be a subset of Ω such that P contains all minimal elements of Ω and such that for every $\sigma \in \Omega$ if $\{\sigma' \in \Omega \mid \sigma' \leq \sigma, \sigma' \neq \sigma\} \subseteq P$ then $\sigma \in P$. Under these conditions $P = \Omega$.

First we show that (iv) holds for all operators $\sigma \in \Omega$ with op(def σ) $= \emptyset$.

Since (iv) clearly holds for all $\sigma \in \Omega$ with $\sigma: w \to s$ and def $\sigma = (w: \emptyset)$, condition (iv) also holds for every $\sigma \in \Omega$ with op(def σ) $= \emptyset$ and $\sigma: \lambda \to s$. If $\sigma: w \to s$, $w: [n] \to S$, $n \neq 0$, and op(def σ) $= \emptyset$ then

$$\mathrm{def}\,\sigma = (w: \{i_1 \overset{e}{=} j_1, \ldots, i_m \overset{e}{=} j_m\})$$

with $i_k, j_k \in [n]$ for every $k = 1, 2, \ldots, m$.

Let $\boldsymbol{a} \in A_w$ and let $f\colon A \to B$ be an injective homomorphism with

$$\boldsymbol{a}f_w \in B_{(w:\{i_1 \overset{e}{=} j_1, \ldots, i_m \overset{e}{=} j_m\})}.$$

Since each $(w\colon i_k \overset{e}{=} j_k)$ for $k \in [m]$ is an existence equation, it follows that $i_k w = j_k w$ for each $k \in [m]$.

Now $\boldsymbol{a}f_w \in B_{\mathrm{def}\sigma}$ implies $(\boldsymbol{a}f_w)\, i_k^B = (\boldsymbol{a}f_w)\, j_k^B$ for each $k \in [m]$. Consequently

$$i_k(\boldsymbol{a}f_w) = j_k(\boldsymbol{a}f_w) \quad \text{or} \quad (i_k\boldsymbol{a})\, f_{i_k w} = (j_k\boldsymbol{a})\, f_{j_k w} \quad \text{for} \quad k \in [m],$$

which implies together with the injectivity of f that $i_k\boldsymbol{a} = j_k\boldsymbol{a}$ for every $k \in [m]$, and therefore $\boldsymbol{a} \in A_{\mathrm{def}\sigma}$.

So far we have proved that (iv) is true for every operator $\sigma \in \Omega$ with $\mathrm{op}(\mathrm{def}\,\sigma) = \emptyset$. For the proof of this condition (i) was not used.

For $\sigma \in \Omega$ with $\mathrm{op}(\mathrm{def}\,\sigma) = \emptyset$ we set

$$\begin{aligned} P(\sigma) = \{\sigma' \in \Omega \mid \sigma' \neq \sigma & \text{ and there is a finite sequence} \\ & \sigma = \sigma_1, \sigma_2, \ldots, \sigma_n = \sigma' \quad \text{with} \\ & \sigma_{i+1} \in \mathrm{op}(\mathrm{def}\,\sigma_i) \text{ for } i = 1, 2, \ldots, n-1\}. \end{aligned}$$

Now we assume that (iv) is true for all $\sigma' \in P(\sigma)$, so that we have to prove that (iv) holds also for σ. Let $\sigma \in \Omega$, $\sigma\colon w \to s$, $f\colon A \to B$ any injective homomorphism, $\boldsymbol{a} \in A_w$ with $\boldsymbol{a}f_w \in A_{\mathrm{def}\sigma}$, and let $(w\colon t_1 \overset{e}{=} t_2) \in \mathrm{def}\,\sigma$ be any existence equation of $\mathrm{def}\,\sigma$.

Then $\mathrm{op}(t_1) \subseteq P(\sigma)$ and $\mathrm{op}(t_2) \subseteq P(\sigma)$. A simple induction shows that (iv)* holds for every term $t \in T(\Sigma, v)$, $v\colon X \to S$, with $\mathrm{op}(t) \subseteq P(\sigma)$ provided (iv) holds for every $\sigma' \in P(\sigma)$.

Now $\boldsymbol{a}f_w \in B_{\mathrm{def}\sigma}$ implies $\boldsymbol{a}f_w \in B_{(w:t_1 \overset{e}{=} t_2)} \supseteq B_{\mathrm{def}\sigma}$. This means that

$$\boldsymbol{a}f_w \in B_{\mathrm{def}t_1}, \quad \boldsymbol{a}f_w \in B_{\mathrm{def}t_2}, \quad \text{and} \quad \boldsymbol{a}f_w t_1^B = \boldsymbol{a}f_w t_2^B.$$

Therefore $\boldsymbol{a} \in A_{\mathrm{def}t_1}$, $\boldsymbol{a} \in A_{\mathrm{def}t_2}$ and analogously to the proof of of (iv) $\to$ (iii) one can see that

$$\boldsymbol{a} \in A_{(w:t_1 \overset{e}{=} t_2)} \quad \text{for every} \quad (w\colon t_1 \overset{e}{=} t_2) \in \mathrm{def}\,\sigma.$$

This means that $\boldsymbol{a} \in A_{\mathrm{def}\sigma}$. In this manner we see that (iv) holds for every $\sigma \in \Omega$.

This concludes the proof of Theorem 2.4.2. ■

Theorem 2.4.2 makes apparent the great importance of the hierarchy condition for a proper theory of partial algebras. We will see that hierarchical equationally partial algebras not only make possible a proper theory but that this concept is general enough to treat a wide spectrum of applications in algebra as well as in computer science.

We call θ-algebras for hierarchical ep-signature *equoids* for short; this is an artificial name.

Definition 2.4.3. Let $\theta = (\Sigma, (\mathrm{def}\,\sigma \mid \sigma \in \Omega))$ be any hep-signature. A θ-equoid A is said to be a *subequoid* of the θ-equoid B if

(i) $A_s \subseteq B_s$ for each $s \in S_\theta$ and

(ii) the S-inclusion $i = (i_s: A_s \to B_s \mid s \in S_\theta)$ with $xi_s = x$ for each $s \in S_\theta$ and $x \in A_s$ forms a homomorphism. ■

Just as in the case of total algebras we can characterize those S-subsets of an equoid A which are carriers of a subequoid as those which are closed with respect to all fundamental operations σ^A.

Corollary 2.4.4. *Let* $\theta = (\Sigma, (\mathrm{def}\,\sigma \mid \sigma \in \Omega))$ *be any hep-signature and* $A \in \mathrm{ALG}(\theta)$ *any equoid. An S-subset* $M \subseteq A$ *is a carrier of a subequoid of A if and only if for every* $\sigma \in \Omega$ *with* $\sigma: w \to s$ *and every* $\boldsymbol{m} \in M_w \cap A_{\mathrm{def}\sigma}$ *it follows that*

$$\boldsymbol{m}\sigma^A \in M_s.$$

Proof. If M is a subequoid of A, then the S-inclusion $i: M \to A$ is a homomorphism, so that for any $\sigma \in \Omega$ with $\sigma: w \to s$, and any $\boldsymbol{m} \in M_w \cap A_{\mathrm{def}\sigma}$ it follows that $\boldsymbol{m}i_w \in A_{\mathrm{def}\sigma}$. As a consequence of Theorem 2.4.2 we obtain

$$\boldsymbol{m} \in M_{\mathrm{def}\sigma}, \quad \text{and} \quad \boldsymbol{m}\sigma^A = \boldsymbol{m}i_w\sigma^A = \boldsymbol{m}\sigma^M f_s = \boldsymbol{m}\sigma^M \in M_s.$$

Secondly, we assume that M is closed with respect to all fundamental operations of A. For every $\sigma \in \Omega$ with $\sigma: w \to s$ we define a partial operation

$$\sigma^M: \mathrm{dom}\,\sigma^M \to M_s \quad \text{by} \quad \mathrm{dom}\,\sigma^M = M_w \cap A_{\mathrm{def}\sigma} \quad \text{and} \quad \boldsymbol{m}\sigma^M = \boldsymbol{m}\sigma^A$$

for every $\boldsymbol{m} \in M_w \cap A_{\mathrm{def}\sigma}$.

Thus we have to show that

$$M = ((M_s \mid s \in S), (\sigma^M \mid \sigma \in \Omega))$$

is not only a partial algebra of type $(S, \alpha: \Omega \to S^* \times S)$ but also an equoid of type θ. One can easily prove inductively that for any S-sorted system of variables $v: X \to S$ the following holds:

(i) If $t \in T(\Sigma, v)_s$ and $\boldsymbol{m} \in \mathrm{dom}\,t^A \cap M_v$ then $\boldsymbol{m}t^A \in M_s$.

(ii) $\mathrm{dom}\,t^M = \mathrm{dom}\,t^M \cap M_v$.

(iii) If we consider the term function $t^M: \mathrm{dom}\,t^M \to M_s$ as a subset of $M_v \times M_s$ with $(\boldsymbol{m}, m) \in t^M$ iff $m = \boldsymbol{m}t^M$ then

$$t^M = t^A \cap (M_v \times M_s).$$

Similarly to the proof of Theorem 2.4.2 we can continue here and show by induction on the hierarchy of operators that

$$\mathrm{dom}\,\sigma^M = M_{\mathrm{def}\sigma}.$$

Therefore as before we set

$$P(\sigma) = \{\sigma' \in \Omega \mid \text{there is a finite sequence } \sigma = \sigma_1, \ldots, \sigma_n = \sigma' \\ \text{with } \sigma_{i+1} \in \mathrm{op}(\mathrm{def}\, \sigma_i) \text{ for } i = 1, 2, \ldots, n-1\}.$$

If $P(\sigma) = \emptyset$ then either $\mathrm{def}\,\sigma = (w{:}\emptyset)$ or

$$\mathrm{def}\,\sigma = (w\colon \{i_1 \overset{e}{=} j_1, \ldots, i_r \overset{e}{=} j_r\})$$

where $i_k, j_k \in [n]$ for every $k = 1, 2, \ldots, r$ and where $\sigma\colon w \to s$ and $w\colon [n] \to S$.
$\mathrm{def}\,\sigma = (w\colon \emptyset)$ implies

$$M_{(w:\emptyset)} = M_w = M_w \cap A_w = M_w \cap A_{(w:\emptyset)} = M_w \cap \mathrm{dom}\,\sigma^A = \mathrm{dom}\,\sigma^M,$$

and in the other case

$$\begin{aligned} M_{\mathrm{def}\sigma} &= \{m \in M_w \mid i_k m = j_k m \text{ for every } k = 1, 2, \ldots, r\} \\ &= M_w \cap \{m \in A_w \mid i_k m = j_k m \text{ for every } k = 1, 2, \ldots, r\} \\ &= M_w \cap A_{\mathrm{def}\sigma} = \mathrm{dom}\,\sigma^M. \end{aligned}$$

Now we turn to the case $P(\sigma) \neq \emptyset$ and show that $M_{\mathrm{def}\sigma} = \mathrm{dom}\,\sigma^M$ holds provided $M_{\mathrm{def}\sigma'} = \mathrm{dom}\,(\sigma')^M$ holds for every $\sigma' \in P(\sigma)$. This assumption leads to $M_{\mathrm{def}t} = \mathrm{dom}\,t^M$ for every $v\colon X \to S$, $s \in S$, $t \in T(\Sigma, v)_s$ with $\mathrm{op}(t) \subseteq P(\sigma)$, so we obtain

$$\begin{aligned} \mathrm{dom}\,\sigma^M &= \mathrm{dom}\,\sigma^A \cap M_w = A_{\mathrm{def}\sigma} \cap M_w \\ &= \cap \{A_{(w:t_1 \overset{e}{=} t_2)} \mid (w\colon t_1 \overset{e}{=} t_2) \in \mathrm{def}\,\sigma\} \cap M_w \\ &= \cap \big(\{A_{\mathrm{def}t_1} \cap A_{\mathrm{def}t_2} \mid (w\colon t_1 \overset{e}{=} t_2) \in \mathrm{def}\,\sigma\} \\ &\qquad \cap \{a \in A_w \mid at_1^A = at_2^A, (w\colon t_1 \overset{e}{=} t_2) \in \mathrm{def}\,\sigma\}\big) \cap M_w \\ &= \{m \in M_w \mid m \in \mathrm{dom}\,t_1^M \cap \mathrm{dom}\,t_2^M, mt_1^M = mt_2^M, \\ &\qquad (w\colon t_1 \overset{e}{=} t_2) \in \mathrm{def}\,\sigma\} = M_{\mathrm{def}\sigma}. \end{aligned}$$

■

The problem we now deal with is the behaviour of subequoids under homomorphisms.

An essential difference from the behaviour of total algebras appears now. Homomorphisms between equoids in general do not preserve subequoids, e.g. the image of a subequoid is in general not the carrier of a subequoid. This fact is very well-known in the theory of small categories. We illustrate this difference by the following homomorphism $f\colon A \to B$ between θ_{CAT}-equoids A and B.

Let A be the θ_{CAT}-equoid with carrier sets

$$A_{\mathrm{Obj}} = \{1, 2, 3, 4\}, \qquad A_{\mathrm{Mor}} = \{e_1, e_2, e_3, e_4, x, y\},$$

and with fundamental operations given by

$$j\ \mathrm{id}^A = e_j \quad \text{for} \quad j = 1, 2, 3, 4,$$
$$x\ \mathrm{source}^A = e_1\ \mathrm{source}^A = e_1\ \mathrm{target}^A = 1,$$
$$x\ \mathrm{target}^A = e_2\ \mathrm{source}^A = e_2\ \mathrm{target}^A = 2,$$
$$y\ \mathrm{source}^A = e_3\ \mathrm{source}^A = e_3\ \mathrm{target}^A = 3,$$
$$y\ \mathrm{target}^A = e_4\ \mathrm{source}^A = e_4\ \mathrm{target}^A = 4,$$
$$\mathrm{dom\ comp}^A = \{(e_1, x), (x, e_2), (e_3, y), (y, e_4), (e_1, e_1), (e_2, e_2), (e_3, e_3), (e_4, e_4)\}$$
$$\mathrm{comp}^A(e_j, e_j) = e_j \quad \text{for} \quad j = 1, 2, 3, 4,$$
$$\mathrm{comp}^A(e_1, x) = \mathrm{comp}^A(x, e_2) = x,$$
$$\mathrm{comp}^A(e_3, y) = \mathrm{comp}^A(y, e_4) = y.$$

There follows a pictorial representation of the category A:

$$1 \xrightarrow{x} 2$$
$$3 \xrightarrow{y} 4$$

Let B be the θ_{CAT}-equoid with carrier sets

$$B_{\mathrm{Obj}} = \{\mathrm{I}, \mathrm{II}, \mathrm{III}\}, \qquad B_{\mathrm{Mor}} = \{E_1, E_2, E_3, X, Y, Z\},$$

and with fundamental operations given by

$$\mathrm{I\ Id}^B = E_1, \qquad \mathrm{II\ Id}^B = E_2, \qquad \mathrm{III\ Id}^B = E_3,$$
$$X\ \mathrm{source}^B = Z\ \mathrm{source}^B = E_1\ \mathrm{source}^B = E_2\ \mathrm{target}^B = \mathrm{I},$$
$$X\ \mathrm{target}^B = E_2\ \mathrm{source}^B = E_2\ \mathrm{target}^B = Y\ \mathrm{source}^B = \mathrm{II},$$
$$Y\ \mathrm{target}^B = Z\ \mathrm{target}^B = E_3\ \mathrm{source}^B = E_3\ \mathrm{target}^B = \mathrm{III},$$
$$\mathrm{dom\ comp}^B = \{(E_1, X), (X, E_2), (E_2, Y), (Y, E_3), (X, Y), (E_1, Z), (Z, E_3), (E_1, E_1), (E_2, E_2), (E_3, E_3)\},$$
$$\mathrm{comp}^B(E_j, E_j) = E_j \quad \text{for} \quad j = 1, 2, 3,$$
$$\mathrm{comp}^B(E_1, X) = \mathrm{comp}^B(X, E_2) = X,$$
$$\mathrm{comp}^B(E_2, Y) = \mathrm{comp}^B(Y, E_3) = Y,$$
$$\mathrm{comp}^B(E_1, Z) = \mathrm{comp}^B(Z, E_3) = \mathrm{comp}^B(X, Y) = Z.$$

The category B can pictorially be represented as follows:

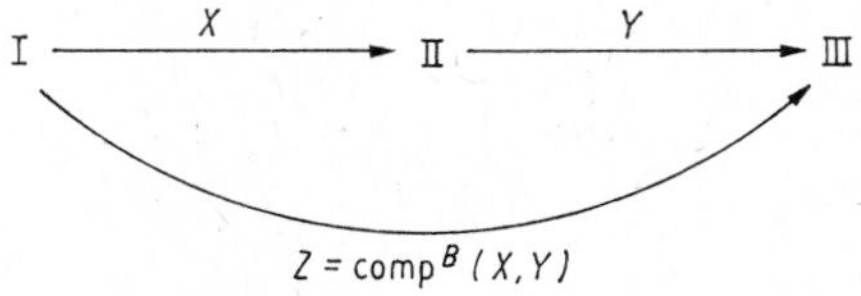

In the pictorial representations of the categories A and B we omitted the identities since they are uniquely represented by the objects.

Let $f: A \to B$ be the homomorphism defined by

$$1f_{\mathrm{Obj}} = \mathrm{I}, \quad 2f_{\mathrm{Obj}} = 3f_{\mathrm{Obj}} = \mathrm{II}, \quad 4f_{\mathrm{Obj}} = \mathrm{III},$$

$$e_1f_{\mathrm{Mor}} = E_1, \quad e_2f_{\mathrm{Mor}} = e_3f_{\mathrm{Mor}} = E_2, \quad e_4f_{\mathrm{Mor}} = E_3,$$

$$xf_{\mathrm{Mor}} = X, \qquad yf_{\mathrm{Mor}} = Y.$$

The image $Af \subseteq B$ is not the carrier of a subequoid of B, since

$$(xf_{\mathrm{Mor}}, yf_{\mathrm{Mor}}) \in \mathrm{dom\ comp}^B$$

and

$$Z = \mathrm{comp}^B(xf_{\mathrm{Mor}}, yf_{\mathrm{Mor}}) \notin Af_{\mathrm{Mor}}.$$

Certainly we can find a simpler counter-example, but as a side-effect we want to demonstrate how laborious it is at this underdeveloped state of the theory to describe concrete equoids. Therefore we will extend to equoids the well-known practice of defining groups by generators and relations. The formalism for doing this is the main point of Chapter 3.

It was not possible to prove that homomorphisms preserve subequoids; however, we can show that homomorphisms reflect subequoids.

Corollary 2.4.5. *Let θ be any hep-signature, $f: A \to B$ any θ-homomorphism and $C \subseteq B$ a subequoid of B. Then the inverse image*

$$Cf^{-1} = (C_sf_s^{-1} \mid s \in S) \quad \textit{where} \quad C_sf_s^{-1} = \{x \in A_s \mid xf_s \in C_s\}$$

for every $s \in S$, is the carrier of a subequoid of A.

Proof. In view of the preceding corollary we have only to show that Cf^{-1} is closed. To this end let be given an operator $\sigma \in \Omega$ with $\sigma: w \to s$ and $\boldsymbol{a} \in (Cf^{-1})_w \cap \mathrm{dom}\,\sigma^A$. Since every homomorphism preserves solutions of domain conditions and since C is a subequoid it follows that

$$\boldsymbol{a}f_w \in \mathrm{dom}\,\sigma^B, \quad \text{so that} \quad \boldsymbol{a}f_w\sigma^B = \boldsymbol{a}\sigma^Af_s \in C_s$$

or in other words

$$\boldsymbol{a}\sigma^A \in C_sf_s^{-1}. \qquad \blacksquare$$

From Corollary 2.4.5 it is easy to show that subequoids are closed under intersection. Thus, we can define the notion of the subequoid generated by an S-subset.

Definition 2.4.6. Let θ be any hep-signature, A any θ-equoid, $X \subseteq A$ any S-subset of the carrier of A, and $f: A \to B$ any homomorphism between θ-equoids.

By $\langle X \rangle$ we denote the smallest subequoid of A containing X. $\langle X \rangle$ is said to be the *subequoid of A generated by X*.

In connection with homomorphisms $\langle Af \rangle$ will be referred to as the *homomorphic image* of A. ■

An immediate consequence of Corollary 2.4.4 and Definition 2.4.6 is the following characterization of the elements of $\langle X \rangle$.

Corollary 2.4.7. *Let θ be any hep-signature, A any θ-equoid and $X \subseteq A$ any S-subset. Then $a \in \langle X \rangle_s$, $s \in S$, if and only if there are w: $[n] \to S$, $t \in T(\Sigma, w)_s$, $\boldsymbol{x} \in X_w$ with $\boldsymbol{x} \in$ dom t^A and $a = \boldsymbol{x}t^A$.*

Proof. Obviously, the S-set of all those elements of A which satisfy the condition of Corollary 2.4.7 contains X and is closed with respect to all fundamental operations of A, so that this S-subset is the carrier of a subequoid containing X. Hence, every element $a \in \langle X \rangle$ satisfies this condition.

By a simple induction one can generalize Corollary 2.4.4 in such a way that any subequoid is closed not only with respect to fundamental operations but also with respect to term functions. ■

For the homomorphic image $\langle Af \rangle$ of a homomorphism f: $A \to B$ one has to pay attention to the fact that the S-subset Af in general is not the carrier of $\langle Af \rangle$ but it is only a generating set.

A further important algebraic construction is the construction of quotient algebras, also called *factor algebras*, by identifying elements. Usually for total algebras, this construction is performed by so-called *congruence relations*, i.e. equivalence relations which are closed with respect to fundamental operations. In the basic theory of partial algebras the concept of a quotient algebra cannot be generalized in a straightforward way. We will not use an element-wise construction of quotient algebras but rather the universal property of the resulting natural homomorphism. At this point we will give only a sketch of the problem, since it will be extensively discussed in Section 3.3.

Definition 2.4.8. Let θ be any hep-signature, A, Q any θ-equoids and ϱ any binary S-relation in the carrier of A.

A homomorphism q: $A \to Q$ is said to be *natural* for ϱ if:

(1) For every $s \in S$ and every $(x, y) \in \varrho_s$, $xq_s = yq_s$ holds.

(2) If f: $A \to B$ is a homomorphism between θ-equoids A and B such that $xf_s = yf_s$ for every $s \in S$ and every $(x, y) \in \varrho_s$, then there exists exactly one homomorphism h: $Q \to B$ with $f = q \circ h$.

If f: $A \to Q$ is natural for an S-relation ϱ in A, then the equoid Q is called a *quotient equoid* of A with respect to ϱ. ■

If for an S-relation ϱ in the carrier of an equoid A natural homomorphisms q_1: $A \to Q_1$, q_2: $A \to Q_2$ exist, then there must be an isomorphism h: $Q_1 \to Q_2$ with $q_2 = q_1 \circ h$, so natural homomorphisms are unique up to isomorphism.

This notion of a natural homomorphism is derived from categorical structure theory, see MacLane (1971), and it coincides with the notion of the natural homomorphism

$$\text{nat}\,\varrho\colon A \to A/\varrho$$

of a congruence relation in a total algebra associating with every element $a \in A$ its congruence class $[a]_\varrho = a(\text{nat}\,\varrho)$.

Here we mention only that a natural homomorphism between equoids is in general not a surjective S-function. The problems of existence and of construction of quotient equoids will be treated in Section 3.3. The problems arise from the fact that by set-theoretic factorizations new solutions of domain conditions may be created in such a way that the value of the fundamental operation for this new solution cannot be reduced to any representative of a congruence class. To illustrate this we return to the θ_{CAT}-equoid A as defined above. If we choose the S-relation ϱ given by

$$\varrho_{\mathrm{Obj}} = \{(2, 3)\}, \qquad \varrho_{\mathrm{Mor}} = \{(e_2, e_3)\}$$

then what we have just described happens.

2.5. hep-varieties and hep-quasi-varieties

Until now we have only generalized the concept of a signature, and we have not yet any means available to describe such properties of operators of sets of operators as associativity, commutativity, or distributivity.

As mentioned in Section 2.2 no kind of validity of term equations is known which is sufficient for describing the basic properties of the set of operations for small categories. Therefore we need other means, which have not a lower expressiveness than existence equations have. Just as for signatures we use the theory of small categories as a guide to discover suitable means. Let us consider the associativity law of small categories, which says that for any morphisms x, y, z the equation

$$\text{comp}^A(\text{comp}^A(x, y), z) = \text{comp}^A(x, \text{comp}^A(y, z))$$

holds, provided $\text{comp}^A(x, y)$, $\text{comp}^A(y, z)$, $\text{comp}^A(\text{comp}^A(x, y), z)$ and $\text{comp}^A(x, \text{comp}^A(y, z))$ are all defined. This kind of conditional equation can be formalized in the following way.

Definition 2.5.1. Let $\Sigma = (S, \alpha\colon \Omega \to S^* \times S)$ be any signature. An *elementary implication*

$$(v\colon G \to t_1 \overset{\mathrm{e}}{=} t_2)$$

is given by a finite S-sorted system of variables $v\colon X \to S$, the *domain* of the elementary implication, by an existence equation $(v\colon t_1 \overset{\mathrm{e}}{=} t_2)$, the *conclusion* of the elementary implication, and by a finite set of existence equations $(v\colon G)$, the *premise* of the elementary implication.

The elementary implication $(v\colon G \to t_1 \overset{e}{=} t_2)$ *holds* in a partial Σ-algebra A if

$$A_{(v:G)} \subseteq A_{(v:t_1 \overset{e}{=} t_2)}.$$

This will be denoted briefly by

$$A \models (v\colon G \to t_1 \overset{e}{=} t_2).$$

An *implication*

$$(v\colon G \to H)$$

is given by an S-sorted system of variables $v\colon X \to S$ and by two sets of existence equations $(v\colon G)$, $(v\colon H)$.

$(v\colon G \to H)$ holds in a partial Σ-algebra A, denoted by

$$A \models (v\colon G \to H), \quad \text{if} \quad A_{(v:G)} \subseteq A_{(v:H)}. \qquad \blacksquare$$

We pay attention to the fact that implications can cause demands of applicability. We illustrate this by means of the hep-signature θ_{CAT} to which we add the elementary implication

$$(w\colon \emptyset \to \mathrm{comp}(\mathrm{comp}(x, y), z) \overset{e}{=} \mathrm{comp}(x, \mathrm{comp}(y, z)))$$

where $w{:}[z] \to \{\mathrm{Obj}, \mathrm{Mor}\}$ with $xw = yw = zw = \mathrm{Mor}$.

If a θ_{CAT}-equoid A satisfies this elementary implication, then

$$A_{(w:\emptyset)} = A_w \subseteq A_{(w:\mathrm{comp}(\mathrm{comp}(x,y),z) \overset{e}{=} \mathrm{comp}(x,\mathrm{comp}(y,z)))}$$

holds, so that every assignment $\boldsymbol{a} \in A_w$ is a solution of the existence equation $(w\colon \mathrm{comp}(\mathrm{comp}(x, y), z) \overset{e}{=} \mathrm{comp}(x, \mathrm{comp}(y, z)))$. This means that $\mathrm{comp}^A(x, y)$ is defined for every $(x, y) \in A_{\mathrm{Mor}} \times A_{\mathrm{Mor}}$. Thus, comp^A is a total binary multiplication. Consequently, we have used the wrong implication to give the associativity law of small categories.

The class of small categories is given by those θ_{CAT}-equoids satisfying the following set $\mathfrak{A}_{\mathrm{CAT}}$ of elementary implications:

$$\begin{aligned}\mathfrak{A}_{\mathrm{CAT}} = \{&(v\colon \emptyset \to \mathrm{source}(\mathrm{id}(e)) \overset{e}{=} e), (v\colon \emptyset \to \mathrm{target}(\mathrm{id}(e)) \overset{e}{=} e),\\ &(v\colon \emptyset \to \mathrm{comp}(\mathrm{id}(\mathrm{source}(x)), x) \overset{e}{=} x),\\ &(v\colon \emptyset \to \mathrm{comp}(x, \mathrm{id}(\mathrm{target}(x))) \overset{e}{=} x),\\ &(v\colon \{\mathrm{target}(x) \overset{e}{=} \mathrm{source}(y)\} \to \mathrm{source}(x) \overset{e}{=} \mathrm{source}(\mathrm{comp}(x, y))),\\ &(v\colon \{\mathrm{target}(x) \overset{e}{=} \mathrm{source}(y)\} \to \mathrm{target}(y) \overset{e}{=} \mathrm{target}(\mathrm{comp}(x, y))),\\ &(v\colon \{\mathrm{target}(x) \overset{e}{=} \mathrm{source}(y), \mathrm{target}(y) \overset{e}{=} \mathrm{source}(z)\}\\ &\quad \to \mathrm{comp}(\mathrm{comp}(x, y), z) \overset{e}{=} \mathrm{comp}(x, \mathrm{comp}(y, z)))\}\end{aligned}$$

where $v\colon \{e, x, y, z\} \to \{\mathrm{Obj}, \mathrm{Mor}\}$ with $ev = \mathrm{Obj}$ and $xv = yv = zv = \mathrm{Mor}$.

The total unary operators $\mathrm{source}(\mathrm{Mor}) \to \mathrm{Obj}$, $\mathrm{target}(\mathrm{Mor}) \to \mathrm{Obj}$ associate with every morphism $x \in A_{\mathrm{Mor}}$ its so-called *source object* $\mathrm{source}^A(x)$

$\in A_{\text{Obj}}$, and *target object* $\text{target}^A(x) \in A_{\text{Obj}}$. The unary total operator $\text{id(Obj)} \to \text{Mor}$ associates with every object $e \in A_{\text{Obj}}$ its so-called *identity morphism* $\text{id}^A(e) \in A_{\text{Mor}}$ which has, according to the first two elementary implications, the given object e as its source and target object. For any morphism $x \in A_{\text{Mor}}$ the morphisms $\text{id}^A(\text{source}^A(x)) \in A_{\text{Mor}}$ and $\text{id}^A(\text{target}^A(x)) \in A_{\text{Mor}}$ are called the *left identity* and *right identity* of x respectively. Their properties are given by the third and fourth elementary implications. The equationally partial binary operator

$$\text{comp}(x\colon \text{Mor}, y\colon \text{Mor } \textbf{iff } \text{target}(x) = \text{source}(y)) \to \text{Mor}$$

associates with any ordered pair of morphisms $(x, y) \in A_{\text{Mor}} \times A_{\text{Mor}}$ a morphism $\text{comp}^A(x, y) \in A_{\text{Mor}}$ provided the target object of x equals the source object of y, i.e. iff (x, y) is a solution of the domain condition. If $\text{comp}^A(x, y)$ is defined, then its source object equals the source object of the first factor, according to the fifth elementary implication, and target objects of $\text{comp}^A(x, y)$ and the second factor are equal, according to the sixth elementary implication. The last elementary implication is the correct form of the associativity law of the partial binary composition.

A typical example of a small category can be obtained if one takes $A_{\text{Obj}} = M$ any set of sets, $A_{\text{Mor}} =$ the set of all functions $f\colon X \to Y$ with $X, Y \in M$. The fundamental operations of the small category are then given by

(a) $\text{id}^A(X) = \text{Id}_X\colon X \to X$ for every $X \in M = A_{\text{Obj}}$,

(b) $\text{source}^A(f\colon X \to Y) = X$, $\text{target}^A(f\colon X \to Y) = Y$ for every $(f\colon X \to Y) \in A_{\text{Mor}}$,

(c) comp^A is the composition of functions.

It is not the aim of this book to give an introduction to the theory of categories, although the author completely agrees with Burstall and Goguen who consider this theory as very useful for computer scientists, see Goguen and Burstall (1982).

As the example of small categories shows, interesting classes of equoids can be described by a hep-signature θ with a set $\mathfrak{A}$ of elementary implications acting as axioms.

But the class of small categories could also be described by an ordinary signature and a modified set of elementary implications, namely by Σ_{CAT} and $\mathfrak{A}^*_{\text{CAT}}$ given by

Σ_{CAT} **is sorts** Obj,Mor
  oprn id(Obj) $\to$ Mor
    source(Mor) $\to$ Obj
    target(Mor) $\to$ Obj
    comp(Mor,Mor) $\circ\!\!\to$ Mor
end

$$\mathfrak{A}^*_{\mathrm{CAT}} = \mathfrak{A}_{\mathrm{CAT}} \cup \{(w\colon \{\mathrm{comp}(x,y) \overset{\mathrm{e}}{=} \mathrm{comp}(x,y)\} \to \mathrm{target}(x) \overset{\mathrm{e}}{=} \mathrm{source}(y)\ ),$$
$$(w\colon \{\mathrm{target}(x) \overset{\mathrm{e}}{=} \mathrm{source}(y)\} \to \mathrm{comp}(x,y) \overset{\mathrm{e}}{=} \mathrm{comp}(x,y)\ )\}.$$

The first additional axiom represents the necessity and the second axiom represents the sufficiency.

Thus, we have at least two formalisms for describing the classes of partial algebras. On the one hand we can use ordinary signatures and sets of elementary implications, and on the other hand we can use hep-signatures and sets of elementary implications.

At first glance the first approach seems to be simpler. However, in this case domain conditions and laws interrelating operators are mixed together and both are represented by elementary implications. Not every set of elementary implications is so well-structured as $\mathfrak{A}^*_{\mathrm{CAT}}$, where we can see which implications give domain conditions and which ones describe operator interrelations.

The second approach is more structured, but not sufficiently so. Domain conditions may be expressed both by hep-signatures and by the set of axioms. Therefore we will restrict the set of axioms in such a way that no additional domain conditions are expressed by the axioms. Then we have a well-structured approach. But it is not a mathematical question which of the two approaches should be favoured. The choice is largely a matter of philosophical taste, together with practical needs.

We prefer the more structured approach where the single parts of the formalism are determined by their functions. Therefore, the hep-signatures are used to describe all required domain conditions and the set $\mathfrak{A}$ of axioms serves to express laws interrelating operations only. Consequently, we are now going to restrict the set of axioms in the announced way.

As preparation we need the concept of *consequences* of a given set of elementary implications. First, we denote the fact that a partial algebra A satisfies all axioms of $\mathfrak{A}$ by

$$A \models \mathfrak{A}.$$

An elementary implication $(v\colon G \to t_1 \overset{\mathrm{e}}{=} t_2)$ is said to be a consequence of the set $\mathfrak{A}$ of elementary implications if for every partial Σ-algebra A the statement $A \models \mathfrak{A}$ implies $A \models (v\colon G \to t_1 \overset{\mathrm{e}}{=} t_2)$. The fact that $\mathfrak{A}$ entails $(v\colon G \to t_1 \overset{\mathrm{e}}{=} t_2)$ will be denoted by

$$\mathfrak{A} \models (v\colon G \to t_1 \overset{\mathrm{e}}{=} t_2).$$

In the same way the corresponding notions are defined for arbitrary implications $(v\colon G \to H)$ and sets of arbitrary implications.

In Section 3.1 we will prove a completeness theorem, so that this semantic definition can be replaced by a syntactic one. In the meantime we can give one way of restricting the set of axioms.

Definition 2.5.2. θ may be any ep-signature. A finite set $\mathfrak{A}$ of elementary implications is said to be *stable* if there exists a partial order $\leq$ on $\mathfrak{A}$ such that for every $(v: G \to t_1 \stackrel{e}{=} t_2) \in \mathfrak{A}$ the following condition is satisfied: $(v: G \to \text{def}\ (t_1 \stackrel{e}{=} t_2))$ is a consequence of

$$\{(u: H \to r_1 \stackrel{e}{=} r_2) \in \mathfrak{A} \mid (u: H \to r_1 \stackrel{e}{=} r_2) < (v: G \to t_1 \stackrel{e}{=} t_2)\},$$

i.e. it is a consequence of all those elementary implications which are less than but not equal to the given one.

If such a partial order on $\mathfrak{A}$ does not exist then $\mathfrak{A}$ is called *unstable*. ■

The set $\mathfrak{A}_{\mathrm{CAT}}$ is an example of a stable set of axioms. To see this let us denote the single elementary implications of $\mathfrak{A}_{\mathrm{CAT}}$ by $a_1, a_2, \ldots, a_6$ respectively. We are going to derive a proper partial order on $\mathfrak{A}_{\mathrm{CAT}}$. Because $\text{def}(v: \text{source}(\text{id(e)}) \stackrel{e}{=} e) = (v: \emptyset)$ and $\text{def}(v: \text{target}(\text{id}(e)) \stackrel{e}{=} e) = (v: \emptyset)$ axioms a_1 and a_2 can be taken as minimal ones. If we consider axioms a_5 and a_6, then

$$\text{def}(v: \text{source}(x) \stackrel{e}{=} \text{source}(\text{comp}(x, y))) = (v: \text{target}(x) \stackrel{e}{=} \text{source}(y))$$
$$= \text{def}(v: \text{target}(y) \stackrel{e}{=} \text{target}(\text{comp}(x, y))),$$

so that a_5 and a_6 can also be taken as minimal ones since

$$\emptyset \models (v: \{\text{target}(x) \stackrel{e}{=} \text{source}(y)\} \to \text{target}(x) \stackrel{e}{=} \text{source}(y)\ ).$$

Owing to the fact that

$$\text{def}(v: \text{comp}(\text{id}(\text{source}(x)), x) \stackrel{e}{=} x) = (v: \text{target}(\text{id}(\text{source}(x))) \stackrel{e}{=} \text{source}(x))$$

the implication $(v: \emptyset \to \text{target}(\text{id}(\text{source}(x))) \stackrel{e}{=} \text{source}(x))$ should be a consequence of a suitable set of minimal axioms. Indeed, this implication is a consequence of $\{a_2\}$ if we replace e by source(x). In the same way one can see that

$$(v: \emptyset \to \text{target}(x) \stackrel{e}{=} \text{source}(\text{id}(\text{target}(x))))$$

is a consequence of $\{a_1\}$, where

$$\text{def}(v: \text{comp}(x, \text{id}(\text{target}(x))) \stackrel{e}{=} x) = (v: \text{target}(x) \stackrel{e}{=} \text{source}(\text{id}(\text{target}(x)))\,).$$

Finally

$$\text{def}(v: \text{comp}(x, \text{comp}(y, z)) = \text{comp}(\text{comp}(x, y, z)))$$
$$= (v: \{\text{target}(x) \stackrel{e}{=} \text{source}(\text{comp}(y, z), \text{target}(\text{comp}(x, y)) \stackrel{e}{=} \text{source}(z)\})$$

holds, so that

$$(v: \{\text{target}(x) \stackrel{e}{=} \text{source}(y), \text{target}(y) \stackrel{e}{=} \text{source}(z)\}$$
$$\to \{\text{target}(x) \stackrel{e}{=} \text{source}(\text{comp}(y, z)), \text{target}(\text{comp}(x, y)) \stackrel{e}{=} \text{source}(z)\})$$

has to be a consequence of the remaining axioms. It is not hard to see that this implication is a consequence of $\{a_5, a_6\}$. Therefore, the partial order is given by $a_1 < a_4$, $a_2 < a_3$, $a_5 < a_7$, $a_6 < a_7$, and the condition of a stable set of axioms is satisfied.

The importance of the concept of stable sets of axioms stems from the following theorem.

Theorem 2.5.3. *Let θ be any ep-signature, $\mathfrak{A}$ any stable set of elementary implications, and $f\colon A \to B$ any injective homomorphism with $B \models \mathfrak{A}$. Then $A \models \mathfrak{A}$ holds, too.*

Proof. By induction we will show that

$$A \models (v\colon G \to t_1 \overset{e}{=} t_2)$$

holds for every elementary implication $(v\colon G \to t_1 \overset{e}{=} t_2)$ of $\mathfrak{A}$. Thus, we assume that $(v\colon G \to t_1 \overset{e}{=} t_2) \in \mathfrak{A}$ and that $A \models (u\colon H \to r_1 \overset{e}{=} r_2)$ holds for every $(u\colon H \to r_1 \overset{e}{=} r_2) \in \mathfrak{A}$ less than but not equal to the given elementary implication. It may be that no such implication exists. This is the case for minimal axioms. Since $\mathfrak{A}$ is stable it follows that

$$A \models \big(v\colon G \to \mathrm{def}(t_1 \overset{e}{=} t_2)\big).$$

Now let $\boldsymbol{a} \in A_{(v:G)}$ and $s \in S$ with $t_1, t_2 \in T(\Sigma, v)_s$. Then

$$\boldsymbol{a} \in \mathrm{dom}\, t_1^A \cap \mathrm{dom}\, t_2^A$$

because

$$A \models \big(v\colon G \to \mathrm{def}(t_1 \overset{e}{=} t_2)\big).$$

If we assume $\boldsymbol{a}t_1^A \neq \boldsymbol{a}t_2^A$, the injectivity of $f\colon A \to B$ yields

$$\boldsymbol{a}t_1^A f_s \neq \boldsymbol{a}t_2^A f_s.$$

Since $f\colon A \to B$ is not only an injective S-function but also a homomorphism, this inequality causes

$$\boldsymbol{a}f_v t_1^B = \boldsymbol{a}t_1^A f_s \neq \boldsymbol{a}t_2^A f_s = \boldsymbol{a}f_v t_2^B.$$

On the other hand $B \models (v\colon G \to t_1 \overset{e}{=} t_2)$ holds because $B \models \mathfrak{A}$.

For $\boldsymbol{a} \in A_{(v:G)}$ this leads to

$$\boldsymbol{a}f_v \in B_{(v:G)} \subseteq B_{(v:t_1 \overset{e}{=} t_2)}.$$

But this contradicts

$$\boldsymbol{a}f_v t_1^B \neq \boldsymbol{a}f_v t_2^B,$$

and hence

$$\boldsymbol{a}t_1^A = \boldsymbol{a}t_2^A$$

is true for every $\boldsymbol{a} \in A_{(v:G)}$. This means $A \models (v\colon G \to t_1 \overset{e}{=} t_2)$, and the proof is finished. ■

The following corollary shows how the concept of stable sets of axioms is related to the good behaviour of homomorphisms between $(\theta, \mathfrak{A})$-algebras. The expression $(\theta, \mathfrak{A})$-algebra denotes θ-algebras satisfying all axioms of $\mathfrak{A}$.

Corollary 2.5.4. *For any ep-signature θ and any stable set $\mathfrak{A}$ of elementary implications the following conditions are equivalent:*

(i) *θ is hierarchical;*

(ii) *Every bijective homomorphism $f: A \to B$ between $(\theta, \mathfrak{A})$-algebras is an isomorphism;*

(iii) *Every injective homomorphism $f: A \to B$ between $(\theta, \mathfrak{A})$-algebras reflects solutions of existence equations;*

(iv) *Every injective homomorphism $f: A \to B$ between $(\theta, \mathfrak{A})$-algebras reflects solutions of domain conditions.*

Proof. If we compare Theorem 2.4.2 and its proof, we see that we have only to extend the proof of (ii) $\Rightarrow$ (i). Corollary 2.5.4 is proved if we can show that the bijective homomorphism $e: F^* \to F$ as constructed in the proof of Theorem 2.4.2 is a homomorphism between $(\theta, \mathfrak{A})$-algebras. Obviously, the θ-algebra F is a $(\theta, \mathfrak{A})$-algebra. Since $e: F^* \to F$ is a bijective homomorphism and $\mathfrak{A}$ is a stable set of axioms, $F^* \models \mathfrak{A}$ holds due to Theorem 2.5.3. ■

The results proved up to this point permit us to suggest hep-signatures with stable sets of elementary implications as a proper formalism to deal with quasi-varieties of partial algebras.

Definition 2.5.5. An ordered pair $T = (\theta, \mathfrak{A})$ consisting of a hep-signature θ and a stable set $\mathfrak{A}$ of elementary implications will be referred to as a *hep-theory*.

The class of all those θ-equoids which satisfies all axioms of $\mathfrak{A}$ is called a *hep-quasi-variety* and will be denoted by $\mathrm{ALG}(\theta, \mathfrak{A})$ or by $\mathrm{ALG}(T)$. ■

If we consider the set $\mathfrak{A}_{\mathrm{CAT}}$ which defines the class of all small categories, then this set of axioms is not only stable but it satisfies a further interesting property. For every elementary implication in $\mathfrak{A}_{\mathrm{CAT}}$ the domain condition of the conclusion is semantically equivalent to the premise, and this equivalence is a consequence of the set of axioms which are less than but not equal to the given implication. This kind of axiom corresponds to our understanding of equations for partial algebras, so we adopt the following convention.

Definition 2.5.6. A stable set of elementary implications $\mathfrak{A}$ is called a set of *hep-equations*, if with respect to the required partial order $\leq$ on $\mathfrak{A}$ the following condition is satisfied:

For every $(v: G \to t_1 \overset{e}{=} t_2) \in \mathfrak{A}$ the implication $(v: \mathrm{def}(t_1 \overset{e}{=} t_2) \to G)$ is a consequence of

$$\{(u: H \to r_1 \overset{e}{=} r_2) \in \mathfrak{A} \mid (u: H \to r_1 \overset{e}{=} r_2) < (v: G \to t_1 \overset{e}{=} t_2)\}.$$

The class of all equoids satisfying all axioms of a set of hep-equations will be called a *hep-variety*. ∎

As mentioned above, the class of all small categories constitutes a hep-variety.

It is not hard to see that hep-varieties and hep-quasi-varieties are closed with respect to subequoids and direct products. But we are not going to study closure properties of hep-varieties and hep-quasi-varieties in detail here or to prove Birkhoff theorems. All that can be derived from corresponding theorems of the basic theory of partial algebras as developed in BURMEISTER (1986).

In Chapter 3 we will also consider model classes that are given by a signature Σ an by a set $\mathfrak{A}$ of elementary implications. ALG($\mathfrak{A}$, Σ) denotes the class of all partial Σ-algebras satisfying each elementary implication in $\mathfrak{A}$. The class of all small categories can then also be described by

$$\mathrm{ALG}(\Sigma_{\mathrm{CAT}}, \mathfrak{A}^*_{\mathrm{CAT}}).$$

We will use the notion of an *algebraic theory* T in both cases if $T = (\theta, \mathfrak{A})$ is a hep-theory and if $T = (\Sigma, \mathfrak{A})$ is given by a signature Σ and by a finite set $\mathfrak{A}$ of elementary implications.

3. Partial algebras defined by generators and relations

A motive for the development of a theory of partial algebras is the extension of the principle of structural induction from total algebras to partial ones. However, the only interesting free partial algebras are those that are free with respect to some restricted class of partial algebras. Therefore we start this chapter with an investigation of sets of elementary implications.

In a similar way as for total algebras free partial algebras can be constructed from sets of terms by factorization. This construction requires a syntactic characterization of consequences of a set of axioms. In Section 3.1 we give a complete set of rules of derivation for elementary implications and prove a soundness theorem.

The main notion of Section 3.2 is the notion of a partial algebra $F(\mathfrak{A}, G, v)$ freely generated by a set of existence equations $(v\colon G)$ with respect to a set $\mathfrak{A}$ of elementary implications serving as axioms. This notion extends the well-known concept of the theory of groups defining a group by generators and relations. In the case of the theory of groups the set of axioms $\mathfrak{A}$ defines the class of groups, the system of variables $v\colon X \to S$ represents the set of generators, and $(v\colon G)$ represents the set of defining relations. This concept comprises the concept of an $\mathfrak{A}$-algebra freely generated by an S-set, if we take for $(v\colon G)$ the empty set of existence equations $(v\colon \emptyset)$. Using these partial algebras $F(\mathfrak{A}, G, v)$ we can finally prove a completeness theorem.

By means of the concept of a partial algebra freely generated by a set of existence equations we can give the construction of partial quotient algebras, of free sums of partial algebras, sometimes called *coproducts*, and of other constructions of colimits. These constructions can be carried out by constructions with defining sets of existence equations. These constructions will be done in Section 3.3.

In the next section we use the construction of partial quotient algebras for an extension of the homomorphism theorem. The homomorphism theorem of total algebras says that a homomorphic image can be reconstructed from the domain and the kernel of a homomorphism. We show that in general the homomorphic image of a homomorphism between equoids cannot be reconstructed in a finite way from the domain and the kernel of the homomorphism. But we will see that the homomorphic image is the directed limit of the chain of iterated quotient equoids starting from the domain of the homomorphism. The case of total algebras is then characterized by the fact that the chain of iterated quotient equoids is of length one.

In the last section of this chapter we deal with the construction of relatively free equoids. This construction becomes the algebraic basis of the last chapter.

3.1. Derivability of elementary implications

A first version of a set of rules of derivation for elementary implications was given in KAPHENGST and REICHEL (1971). A completeness theorem for first order languages on homogeneous partial algebras was given by THIELE (1966); however, this was not noticed by algebraists. It is worth mentioning that rules of derivation of homogeneous algebras cannot be extended directly to many-sorted algebras. This problem is thoroughly discussed in GOGUEN and MESEGUER (1981).

We will give a complete set of rules of derivation that allows us to describe syntactically that an existence equation $(v\colon t_1 \overset{e}{=} t_2)$ is a consequence of a set of existence equations $(v\colon G)$ with respect to a given set $\mathfrak{A}$ of axioms. The system of rules of derivation is a very condensed one since there is only one rule that can be applied iteratively. This system was published first in KAPHENGST (1981).

In principle there is not a too big difference between the system of inference rules of GOGUEN and MESEGUER (1981) and the following one. Both systems use binding of the variables to equations and conditional equations similar to the λ-abstraction in combinatory logic. The main difference results from partiality and it becomes explicitly visible in condition (b) of derivation rule DIII in Definition 3.1.2.

For the sake of simplicity we are working with partial Σ-algebras for any signature $\Sigma = (S, \alpha\colon \Omega \to S^* \times S)$. The concept of hep-signatures is not of importance for this section.

Definition 3.1.1. Let $\Sigma = (S, \alpha\colon \Omega \to S^* \times S)$ be any signature. For every $s \in S$, $v_s\colon \{x, y, z\} \to S$ may be the S-sorted system of variables with $xv_s = yv_s = zv_s = s$. Then the following elementary implications are called *tautological:*

$$\text{REF}_s\colon \quad (v_s\colon \emptyset \to x \overset{e}{=} x), \qquad s \in S,$$

$$\text{SYM}_s\colon \quad (v_s\colon x \overset{e}{=} y \to y \overset{e}{=} x), \qquad s \in S,$$

$$\text{TRA}_s\colon \quad (v_s\colon x \overset{e}{=} y, y \overset{e}{=} z \to x \overset{e}{=} z), \qquad s \in S.$$

For every $\sigma \in \Omega$ with $\sigma\colon w \to s$ and $w\colon [n] \to S$ and for every $j \in [n]$ we define the S-sorted system of variables

$$v_{\sigma,j}\colon \{x_1, \ldots, x_n, y, z\} \to S$$

by $x_i v_{\sigma,j} = iw$ for every $i \in [n]$, and $yv_{\sigma,j} = jw$, $zv_{\sigma,j} = s$. With this S-sorted system of variables we define further tautological elementary implications:

$$\text{SUB}_{\sigma,j}\colon \bigl(v_{\sigma,j}\colon x_j \overset{e}{=} y, z \overset{e}{=} (x_1, \ldots, x_n)\,\sigma \to z \overset{e}{=} (x_1, \ldots, x_{j-1}, y, x_{j+1}, \ldots, x_n)\,\sigma\bigr).$$

Finally we call

$$\mathrm{TAUT}(\Sigma) = \{\mathrm{REF}_s \mid s \in S\} \cup \{\mathrm{SYM}_s \mid s \in S\} \cup \{\mathrm{TRA}_s \mid s \in S\}$$
$$\cup \{\mathrm{SUB}_{\sigma,j} \mid \sigma \in \Omega,\ \sigma\colon w \to s,\ w\colon [n] \to S,\ j \in [n]\}$$

the set of tautological elementary implications of Σ. ■

These elementary implications are called *tautological* since

$$\emptyset \models (v\colon G \to t_1 \overset{e}{=} t_2)$$

holds for every

$$(v\colon G \to t_1 \overset{e}{=} t_2) \in \mathrm{TAUT}(\Sigma).$$

Definition 3.1.2. An existence equation $(v\colon t_1 \overset{e}{=} t_2)$ with $v\colon \{x_1, \ldots, x_n\} \to S$ is said to be $\mathfrak{A}$-*derivable* from a finite set of existence equations $(v\colon G)$, denoted by

$$\mathfrak{A} \vdash (v\colon G \to t_1 \overset{e}{=} t_2),$$

where $\mathfrak{A}$ is any set of elementary implications, if it is derivable using only the following rules of derivation DI, DII, DIII:

DI: $\mathfrak{A} \vdash (v\colon G \to t_1 \overset{e}{=} t_2)$ for every $(v\colon t_1 \overset{e}{=} t_2) \in (v\colon G)$;

DII: $\mathfrak{A} \vdash (v\colon G \to x \overset{e}{=} x)$ for every $x \in X$ with $v\colon X \to S$;

DIII: If

(a) $(w\colon H \to t_1 \overset{e}{=} t_2) \in \mathrm{TAUT}(\Sigma) \cup \mathfrak{A}$ with $w\colon Y \to S$,

(b) $\boldsymbol{t} \in T(\Sigma, v)_w$ such that for every $y \in Y$ an existence equation $(v\colon r_1 \overset{e}{=} r_2)$ exists so that
$\mathfrak{A} \vdash (v\colon G \to r_1 \overset{e}{=} r_2)$ and $y\boldsymbol{t}$ is a subterm of r_1 or r_2, and if

(c) $\mathfrak{A} \vdash (v\colon G \to p_1\tilde{\boldsymbol{t}} \overset{e}{=} p_2\tilde{\boldsymbol{t}})$ holds for every $(w\colon p_1 \overset{e}{=} p_2) \in (w\colon H)$ then it follows that

$$\mathfrak{A} \vdash (v\colon G \to t_1\tilde{\boldsymbol{t}} \overset{e}{=} t_2\tilde{\boldsymbol{t}}).$$

$\mathrm{IMP}(\mathfrak{A})$ denotes the set of all elementary implications $(v\colon G \to t_1 \overset{e}{=} t_2)$ with $\mathfrak{A} \vdash (v\colon G \to t_1 \overset{e}{=} t_2)$. ■

To improve the understanding of this definition we give another description of the set $\mathrm{IMP}(\mathfrak{A})$. The rule of derivation DIII is applicable only if there are any $\mathfrak{A}$-derivable elementary implications satisfying (a), (b), and (c). On the other hand, by means of DI and DII one can derive elementary implications without any prerequisites. Therefore we denote by

$$\mathrm{IMP}_0(\mathfrak{A})$$

the set of all those elementary implications that are $\mathfrak{A}$-derivable only by means of DI and DII. This means

$$\mathrm{IMP}_0(\mathfrak{A}) = \{(v\colon G \to t_1 \overset{e}{=} t_2) \mid (v\colon t_1 \overset{e}{=} t_2) \in (v\colon G)\}$$
$$\cup \{(v\colon G \to x \overset{e}{=} x) \mid v\colon X \to S, x \in X\}.$$

Since DIII is the only rule of derivation that can be applied iteratively, we set

$\mathrm{IMP}_{i+1}(\mathfrak{A}) =$ $\mathrm{IMP}_i(\mathfrak{A})$ joined with the set of all those elementary implications $(v\colon G \to t_1\tilde{\boldsymbol{t}} \overset{e}{=} t_2\tilde{\boldsymbol{t}})$ that are $\mathfrak{A}$-derivable by DIII in such a way that in (b) for every $y \in Y$ an existence equation $(v\colon r_1 \overset{e}{=} r_2)$ exists with $(v\colon G \to r_1 \overset{e}{=} r_2) \in \mathrm{IMP}_i(\mathfrak{A})$ and $y\boldsymbol{t}$ is a subterm of r_1 or r_2, and that with respect to (c) $(v\colon G \to p_1\tilde{\boldsymbol{t}} \overset{e}{=} p_2\tilde{\boldsymbol{t}}) \in \mathrm{IMP}_i(\mathfrak{A})$ holds for every $(w\colon p_1 \overset{e}{=} p_2) \in (w\colon H)$.

Thus, $\mathrm{IMP}_i(\mathfrak{A})$ is the set of all elementary implications which are $\mathfrak{A}$-derivable by i-fold application of the rule DIII. This leads to

$$\mathrm{IMP}(\mathfrak{A}) = \bigcup_{i=0}^{\infty} \mathrm{IMP}_i(\mathfrak{A}).$$

Naturally, everybody expects that every tautological elementary implication and every elementary implication of $\mathfrak{A}$ is $\mathfrak{A}$-derivable. But this does not follows immediately from Definition 3.1.2. For this reason we prove the following corollary.

Corollary 3.1.3. *For every set of elementary implications*

$$\mathfrak{A} \cup \mathrm{TAUT}(\Sigma) \subseteq \mathrm{IMP}_1(\mathfrak{A}) \subseteq \mathrm{IMP}(\mathfrak{A})$$

holds.

Proof. According to the preceding consideration we have to show

$$\mathfrak{A} \cup \mathrm{TAUT}(\Sigma) \subseteq \mathrm{IMP}_1(\mathfrak{A}).$$

Therefore, let $(w\colon H \to t_1 \overset{e}{=} t_2)$ be in $\mathrm{TAUT}(\Sigma) \cup \mathfrak{A}$ with $w\colon Y \to S$, and let $\boldsymbol{t} \in T(\Sigma, w)_w$ with $y\boldsymbol{t} = y$ for every $y \in Y$. Because of DII, $\mathfrak{A} \vdash (w\colon H \to y \overset{e}{=} y)$ holds for every $y \in Y$ and even $(w\colon H \to y \overset{e}{=} y) \in \mathrm{IMP}_0(\mathfrak{A})$ for every $y \in Y$. Evidently, $y\boldsymbol{t} = y$ is a subterm of y so that $\boldsymbol{t} \in T(\Sigma, w)_w$ satisfies condition (b) of DIII. The rule of derivation DI implies

$$\mathfrak{A} \vdash (w\colon H \to p_1 \overset{e}{=} p_2) \quad \text{for every} \quad (w\colon p_1 \overset{e}{=} p_2) \in (w\colon H),$$

and so

$$(w\colon H \to p_1 \overset{e}{=} p_2) \in \mathrm{IMP}_0(\mathfrak{A}).$$

Since $\tilde{\boldsymbol{t}}$ is the identity of $T(\Sigma, w)$ we obtain

$$\mathfrak{A} \vdash (w: H \to t_1 \overset{e}{=} t_2) \quad \text{and} \quad (w: H \to t_1 \overset{e}{=} t_2) \in \mathrm{IMP}_1(\mathfrak{A})$$

by rule DIII. ■

Another useful property of $\mathfrak{A}$-derivability is given by the following corollary.

Corollary 3.1.4. *If* $\mathfrak{A} \vdash (v: G \to t_1 \overset{e}{=} t_2)$, *and if* $r \in \mathrm{G}(\Sigma, v)_s$, $s \in S$, *is a subterm of* t_1 *or* t_2, *then* $\mathfrak{A} \vdash (v: G \to r \overset{e}{=} r)$.

Proof. If we take $(v_s: \emptyset \to x \overset{e}{=} x) \in \mathrm{TAUT}(\Sigma)$ and $\boldsymbol{t}: \{x, y, z\} \to T(\Sigma, v)_s$ with $x\boldsymbol{t} = y\boldsymbol{t} = z\boldsymbol{t} = r$, then conditions (b) and (c) are satisfied with respect to $(v: G)$, so we obtain

$$\mathfrak{A} \vdash (v: G \to x\tilde{\boldsymbol{t}} \overset{e}{=} x\tilde{\boldsymbol{t}}) \quad \text{or} \quad \mathfrak{A} \vdash (v: G \to r \overset{e}{=} r)$$

by DIII. ■

Next we formulate the *generalized substitution rule* which will be referred to by GSR.

Corollary 3.1.5. *If* $v: X \to S$, $w: Y \to S$ *are finite S-sorted systems of variables, if* $\boldsymbol{t}_1, \boldsymbol{t}_2 \in T(\Sigma, v)_w$ *are assignments, and if* $t \in T(\Sigma, w)_s$, $s \in S$, *then from*

$$\mathfrak{A} \vdash (v: G \to y\tilde{\boldsymbol{t}}_1 \overset{e}{=} y\tilde{\boldsymbol{t}}_2) \quad \textit{for every} \quad y \in Y,$$

and

$$\mathfrak{A} \vdash (v: G \to t\tilde{\boldsymbol{t}}_1 \overset{e}{=} t\tilde{\boldsymbol{t}}_1)$$

follows

$$\mathfrak{A} \vdash (v: G \to t\tilde{\boldsymbol{t}}_1 \overset{e}{=} t\tilde{\boldsymbol{t}}_2).$$

Proof. GSR will be proved by induction on the structure of $t \in T(\Sigma, w)$.

If $t = y \in T(\Sigma, w)_{yw}$ or $t = \sigma^{T(\Sigma,w)}$ with $\sigma: \lambda \to s$, i.e. if $\sigma \in \Omega$ is a constant, then the conclusion of GSR coincides with one of the premises.

Corresponding to the inductive proof we assume that $t = \boldsymbol{p}\sigma^{T(\Sigma,w)}$ with $\sigma: u \to s$, $u: [n] \to S$, $\boldsymbol{p} \in T(\Sigma, w)_u$, and that GSR is applicable to $i\boldsymbol{p}$ for every $i \in [n]$.

Since

$$(\boldsymbol{p}\sigma^{T(\Sigma,w)})\, \tilde{\boldsymbol{t}}_1 = \boldsymbol{p}\tilde{\boldsymbol{t}}_1\sigma^{T(\Sigma,v)}$$

the premises of GSR yield

$$\mathfrak{A} \vdash (v: G \to \boldsymbol{p}\tilde{\boldsymbol{t}}_1\sigma^{T(\Sigma,v)} \overset{e}{=} \boldsymbol{p}\tilde{\boldsymbol{t}}_1\sigma^{T(\Sigma,v)})$$

and

$$\mathfrak{A} \vdash (v: G \to y\tilde{\boldsymbol{t}}_1 \overset{e}{=} y\tilde{\boldsymbol{t}}_2) \quad \text{for every} \quad y \in Y.$$

Now we take the tautological elementary implication $\mathrm{SUB}_{\sigma,1}$, defined in Definition 3.1.1, and

$$\boldsymbol{r} \in T(\Sigma, v)_{v_{\sigma,1}} \quad \text{with} \quad x_j\boldsymbol{r} = (j\boldsymbol{p})\, \tilde{\boldsymbol{t}}_1 \quad \text{for} \quad j \in [n],$$

with

$$z\boldsymbol{r} = \boldsymbol{p}\tilde{\boldsymbol{t}}_1\sigma^{T(\Sigma,v)}, \qquad y\boldsymbol{r} = (1\boldsymbol{p})\,\tilde{\boldsymbol{t}}_2,$$

and we check conditions (b) and (c) of DIII with respect to $\mathrm{SUB}_{\sigma,1}$,

$$\boldsymbol{r} \in T(\Sigma, v)_{v_{\sigma,1}}$$

and with respect to $(v\colon G)$. Since GSR is applicable to $\mathbf{1}\boldsymbol{p}$,

$$\mathfrak{A} \vdash \left(v\colon G \to (1\boldsymbol{p})\,\tilde{\boldsymbol{t}}_1 \overset{\mathrm{e}}{=} (1\boldsymbol{p})\,\tilde{\boldsymbol{t}}_2\right)$$

follows. This together with

$$\mathfrak{A} \vdash (v\colon G \to \boldsymbol{p}\tilde{\boldsymbol{t}}_1\sigma^{T(\Sigma,v)} \overset{\mathrm{e}}{=} \boldsymbol{p}\tilde{\boldsymbol{t}}_1\sigma^{T(\Sigma,v)})$$

guarantees that for every $x \in \{y, z, x_1, \ldots, x_n\}$ there is an existence equation $(v\colon q_1 \overset{\mathrm{e}}{=} q_2)$ with

$$\mathfrak{A} \vdash (v\colon G \to q_1 \overset{\mathrm{e}}{=} q_2)$$

so that $x\boldsymbol{r}$ is a subterm of q_1 or q_2.

Condition (c) of DIII is satisfied because

$$\mathfrak{A} \vdash \left(v\colon G \to (1\boldsymbol{p})\,\tilde{\boldsymbol{t}}_1 \overset{\mathrm{e}}{=} (1\boldsymbol{p})\,\tilde{\boldsymbol{t}}_2\right),$$

$$\mathfrak{A} \vdash (v\colon G \to \boldsymbol{p}\tilde{\boldsymbol{t}}_1\sigma^{T(\Sigma,v)} \overset{\mathrm{e}}{=} \boldsymbol{p}\tilde{\boldsymbol{t}}_1\sigma^{T(\Sigma,v)}),$$

and

$$\left((x_1, \ldots, x_n)\,\sigma^{T(\Sigma,v_{\sigma,1})}\right)\tilde{\boldsymbol{r}} = (x_1\boldsymbol{r}, \ldots, x_n\boldsymbol{r})\,\sigma^{T(\Sigma,v)} = \boldsymbol{p}\tilde{\boldsymbol{t}}_1\sigma^{T(\Sigma,v)}.$$

By means of DIII we get

$$\mathfrak{A} \vdash \left((v\colon G \to z\boldsymbol{r} \overset{\mathrm{e}}{=} (y, x_2, \ldots, x_n)\,\sigma^{T(\Sigma,v_{\sigma,1})})\,\tilde{\boldsymbol{r}}\right)$$

and so

$$\mathfrak{A} \vdash \left(v\colon G \to \boldsymbol{p}\tilde{\boldsymbol{t}}_1\sigma^{T(\Sigma,v)} \overset{\mathrm{e}}{=} \left((1\boldsymbol{p})\,\tilde{\boldsymbol{t}}_1, \ldots, (n\boldsymbol{p})\,\tilde{\boldsymbol{t}}_1\right)\sigma^{T(\Sigma,v)}\right).$$

By an $(n - 1)$-fold iteration of the preceding step one can finally prove

$$\mathfrak{A} \vdash (v\colon G \to \boldsymbol{t}\tilde{\boldsymbol{t}}_1 \overset{\mathrm{e}}{=} \boldsymbol{t}\tilde{\boldsymbol{t}}_2). \qquad \blacksquare$$

Finally we want to extend the rule of derivation DIII in such a way that in the premise (a) not only axioms or tautological elementary implications are allowed but arbitrary $\mathfrak{A}$-derivable elementary implications.

Corollary 3.1.6. IMP($\mathfrak{A}$) *is closed with respect to the following extended rule of derivation EDIII:*
If

(a) $\mathfrak{A} \vdash (w\colon H \to t_1 \overset{\mathrm{e}}{=} t_2)$, $w\colon Y \to S$,

(b) $\boldsymbol{t} \in T(\Sigma, v)_w$ *such that for every* $y \in Y$ *there is an existence equation* $(v\colon r_1 \overset{\mathrm{e}}{=} r_2)$ *with* $\mathfrak{A} \vdash (v\colon G \to r_1 \overset{\mathrm{e}}{=} r_2)$ *so that* $y\boldsymbol{t}$ *is a subterm of* r_1 *or* r_2, *and if*

(c) $\mathfrak{A} \vdash (v\colon G \to p_1\tilde{\boldsymbol{t}} \overset{e}{=} p_2\boldsymbol{t})$ *for every* $(w\colon p_1 \overset{e}{=} p_2) \in (w\colon H)$, *then*

$$\mathfrak{A} \vdash (v\colon G \to t_1\tilde{\boldsymbol{t}} \overset{e}{=} t_2\boldsymbol{t}).$$

Proof. Corollary 3.1.6 will be proved by use of the inductive representation

$$\mathrm{IMP}(\mathfrak{A}) = \bigcup_{i=0}^{\infty} \mathrm{IMP}_i(\mathfrak{A}).$$

First we show that EDIII is correct if in (a) an elementary implication of $\mathrm{IMP}_0(\mathfrak{A})$ is used. Let

$$(w\colon H \to t_1 \overset{e}{=} t_2) \in \mathrm{IMP}_0(\mathfrak{A}) \quad \text{with} \quad w\colon Y \to S, \quad \boldsymbol{t} \in T(\Sigma, v)_w,$$

and let $(v\colon G)$ be a set of existence equations such that (b) and (c) are satisfied. On account of the definition of $\mathrm{IMP}_0(\mathfrak{A})$ two cases are to be considered. One case is given by $(w\colon t_1 \overset{e}{=} t_2) \in (w\colon H)$ and the other one by $t_1 = t_2 = y_0 \in Y$.

If $(w\colon t_1 \overset{e}{=} t_2) \in (w\colon H)$, then condition (c) implies

$$\mathfrak{A} \vdash (v\colon G \to p_1\tilde{\boldsymbol{t}} \overset{e}{=} p_2\tilde{\boldsymbol{t}}) \quad \text{for all} \quad (w\colon p_1 \overset{e}{=} p_2) \in (w\colon H),$$

and so

$$\mathfrak{A} \vdash (v\colon G \to t_1\tilde{\boldsymbol{t}} \overset{e}{=} t_2\tilde{\boldsymbol{t}}).$$

Now let $t_1 = t_2 = y_0 \in Y$ and $s = y_0w$. Then condition (b) together with Corollary 3.1.4 yields

$$\mathfrak{A} \vdash (v\colon G \to y\boldsymbol{t} \overset{e}{=} y\boldsymbol{t}) \quad \text{for every} \quad y \in Y.$$

This leads to

$$\mathfrak{A} \vdash (v\colon G \to t_1\tilde{\boldsymbol{t}} \overset{e}{=} t_2\tilde{\boldsymbol{t}})$$

because

$$t_1\tilde{\boldsymbol{t}} = y_0\boldsymbol{t} = y_0\boldsymbol{t} = t_2\tilde{\boldsymbol{t}}.$$

Now, according to the induction, we assume that EDIII produces $\mathfrak{A}$-derivable elementary implications, provided in (a) an elementary implication of $\mathrm{IMP}_i(\mathfrak{A})$, $i \geq 0$, is used. We have to show by means of this assumption that EDIII also produces $\mathfrak{A}$-derivable elementary implications if in (a) an elementary implication of $\mathrm{IMP}_{i+1}(\mathfrak{A})$ is used.

Let

$$(w\colon H \to t_1 \overset{e}{=} t_2) \in \mathrm{IMP}_{i+1}(\mathfrak{A}), \quad w\colon Y \to S, \quad \boldsymbol{t} \in T(\Sigma, v)_w,$$

$v\colon X \to S$, and $(v\colon G)$ be given such that (b) and (c) of EDIII are satisfied. Because

$$(w\colon H \to t_1 \overset{e}{=} t_2) \in \mathrm{IMP}_{i+1}(\mathfrak{A})$$

there is an elementary implication

$$(u\colon H^* \to q_1 \overset{e}{=} q_2) \in \mathrm{TAUT}(\Sigma) \cup \mathfrak{A}, \qquad u\colon Z \to S,$$

and there is an assignment $\boldsymbol{r} \in T(\Sigma, w)_u$ such that (b) and (c) of DIII are satisfied with respect to $(w\colon H)$, and furthermore $t_1 = q_1\tilde{\boldsymbol{r}}$, $t_2 = q_2\tilde{\boldsymbol{r}}$.

Thus, we have an implication

$$(u\colon H^* \to q_1 \overset{\mathrm{e}}{=} q_2) \in \mathrm{TAUT}(\Sigma) \cup \mathfrak{A}$$

and an assignment $\boldsymbol{r}\tilde{\boldsymbol{t}} \in T(\Sigma, v)_u$ which satisfy (b) and (c) of DIII with respect to $(v\colon G)$, as will be proved in the following.

First we consider condition (b):

For every $z \in Z$ there is an existence equation $(w\colon p_1 \overset{\mathrm{e}}{=} p_2)$ with

$$(w\colon H \to p_1 \overset{\mathrm{e}}{=} p_2) \in \mathrm{IMP}_i(\mathfrak{A})$$

so that $z\boldsymbol{r}$ is a subterm of p_1 or p_2. According to our assumption we can apply EDIII to $(w\colon H \to p_1 \overset{\mathrm{e}}{=} p_2)$. So for every $z \in Z$ there is an existence equation $(v\colon p_1\tilde{\boldsymbol{t}} = p_2\tilde{\boldsymbol{t}})$ with

$$\mathfrak{A} \vdash (v\colon G \to p_1\tilde{\boldsymbol{t}} = p_2\tilde{\boldsymbol{t}})$$

so that $(z\boldsymbol{r})\,\tilde{\boldsymbol{t}}$ is a subterm of $p_1\tilde{\boldsymbol{t}}$ or $p_2\tilde{\boldsymbol{t}}$, because $z\boldsymbol{r}$ is a subterm of p_1 or p_2, and

$$\tilde{\boldsymbol{t}}\colon T(\Sigma, w) \to T(\Sigma, v)$$

preserves the subterm relation. Consequently, (b) is satisfied.

Next we consider condition (c), i.e. we have to prove that

$$\mathfrak{A} \vdash \big(v\colon G \to p_1(\widetilde{\boldsymbol{r}\tilde{\boldsymbol{t}}}) \overset{\mathrm{e}}{=} p_2(\widetilde{\boldsymbol{r}\tilde{\boldsymbol{t}}})\big)$$

holds for every $(u\colon p_1 \overset{\mathrm{e}}{=} p_2) \in (u\colon H^*)$. The required properties of $\boldsymbol{r} \in T(\Sigma, w)$ and

$$(u\colon H^* \to q_1 \overset{\mathrm{e}}{=} q_2) \in \mathrm{TAUT}(\Sigma) \cup \mathfrak{A}$$

yield

$$(w\colon H \to p_1\tilde{\boldsymbol{r}} \overset{\mathrm{e}}{=} p_2\tilde{\boldsymbol{r}}) \in \mathrm{IMP}_i(\mathfrak{A}) \quad \text{for every} \quad (u\colon p_1 \overset{\mathrm{e}}{=} p_2) \in (u\colon H^*).$$

Hence we can apply EDIII to each of these elementary implications and we obtain

$$\mathfrak{A} \vdash \big(v\colon G \to (p_1\tilde{\boldsymbol{r}})\,\tilde{\boldsymbol{t}} \overset{\mathrm{e}}{=} (p_2\tilde{\boldsymbol{r}})\,\tilde{\boldsymbol{t}}\big)$$

for every $(u\colon p_1 \overset{\mathrm{e}}{=} p_2) \in (u\colon H^*)$. Condition (c) is also satisfied because

$$p_1(\widetilde{\boldsymbol{r}\tilde{\boldsymbol{t}}}) = (p_1\tilde{\boldsymbol{r}})\,\tilde{\boldsymbol{t}} \quad \text{and} \quad p_2(\widetilde{\boldsymbol{r}\tilde{\boldsymbol{t}}}) = (p_2\tilde{\boldsymbol{r}})\,\tilde{\boldsymbol{t}}.$$

Now we can apply DIII to $(u\colon H^* \to q_1 \overset{\mathrm{e}}{=} q_2) \in \mathrm{TAUT}(\Sigma) \cup \mathfrak{A}$, $\boldsymbol{r}\tilde{\boldsymbol{t}} \in T(\Sigma, v)_u$ and $(v\colon G)$, and we obtain

$$\mathfrak{A} \vdash \big(v\colon G \to q_1(\widetilde{\boldsymbol{r}\tilde{\boldsymbol{t}}}) \overset{\mathrm{e}}{=} q_2(\widetilde{\boldsymbol{r}\tilde{\boldsymbol{t}}})\big).$$

Because

$$q_1(\widetilde{\boldsymbol{r}\tilde{\boldsymbol{t}}}) = (q_1\tilde{\boldsymbol{r}})\,\tilde{\boldsymbol{t}} = t_1\tilde{\boldsymbol{t}} \quad \text{and} \quad q_2(\widetilde{\boldsymbol{r}\tilde{\boldsymbol{t}}}) = t_2\tilde{\boldsymbol{t}}$$

we have proved

$$(v\colon G \to t_1\tilde{\boldsymbol{t}} \stackrel{\mathrm{e}}{=} t_2\tilde{\boldsymbol{t}}) \in \mathrm{IMP}(\mathfrak{A}). \qquad \blacksquare$$

The preceding corollaries and their proofs may give a better understanding of the rather compact system of inference rules in Definition 3.1.2. By the following example of $\mathfrak{A}$-deriving an elementary implication we will strengthen this understanding.

The system $\mathfrak{A}_0$ of axioms may be given by the semantics definition of the while-construction, i.e.

$$\mathfrak{A}_0 = \{(w_0\colon \{p(y) = \text{false}, f(y) = f(y)\} \to \text{while}(y) = f(y)),$$
$$w_0\colon \{p(y) = \text{true}, \text{while}(g(y)) = \text{while}(g(y))\} \to \text{while}(y) = \text{while}(g(y))\},$$

with $w_0\colon \{y\} \to \{M\}$, and we want to $\mathfrak{A}_0$-derive the elementary implication

$$(v_0\colon \{p(x) = \text{true}, p(g(x)) = \text{true}, \text{while}(g(g(x))) = \text{while}(g(g(x)))\}$$
$$\to \text{while}(x) = \text{while}(g(g(x))))$$

with $v_0\colon \{x\} \to \{M\}$, see specification WHILE in Section 1.2.

For an application of DIII we have to fix single parameters. Since the intended implication will be proved by a repeated application of DIII we set for the first application:

$$(w\colon H \to t_1 \stackrel{\mathrm{e}}{=} t_2) = (w_0\colon \{p(y) \stackrel{\mathrm{e}}{=} \text{true}, \text{while}(g(y)) \stackrel{\mathrm{e}}{=} \text{while}(g(y))\}$$
$$\to \text{while}(y) \stackrel{\mathrm{e}}{=} \text{while}(g(y))),$$

i.e. $w = w_0$, $\boldsymbol{t} \in T(\Sigma, v_0)_{w_0}$ is given by $y\boldsymbol{t} \stackrel{\mathrm{e}}{=} g(x)$, and

$$(v\colon G) = (v_0\colon \{p(x) \stackrel{\mathrm{e}}{=} \text{true}, p(g(x)) \stackrel{\mathrm{e}}{=} \text{true}, \text{while}(g(g(x))) \stackrel{\mathrm{e}}{=} \text{while}(g(g(x)))\}).$$

Now we have to prove that there is an existence equation $(v_0\colon r_1 \stackrel{\mathrm{e}}{=} r_2)$ such that $\mathfrak{A} \vdash (v_0\colon G \to r_1 \stackrel{\mathrm{e}}{=} r_2)$ and $y\boldsymbol{t} = g(x)$ is a subterm of r_1 or r_2. If we take $(v_0\colon p(g(x)) \stackrel{\mathrm{e}}{=} \text{true}) \in (v_0\colon G)$ for that equation, then $\mathfrak{A}_0 \vdash (v_0\colon G \to p(g(x)) \stackrel{\mathrm{e}}{=} \text{true})$, due to DI and the fact that $g(x)$ is a subterm of $p(g(x))$. Next we have to prove $\mathfrak{A}_0 \vdash (v_0\colon G \to p_1\boldsymbol{t} \stackrel{\mathrm{e}}{=} p_2\boldsymbol{t})$ for all $(w_0\colon p_1 \stackrel{\mathrm{e}}{=} p_2) \in (w\colon H)$, i.e. we have to prove

$$\mathfrak{A}_0 \vdash (w_0\colon G \to p(g(x)) \stackrel{\mathrm{e}}{=} \text{true})$$

and

$$\mathfrak{A}_0 \vdash (w_0\colon G \to \text{while}(g(g(x))) \stackrel{\mathrm{e}}{=} \text{while}(g(g(x)))).$$

Evidently, both statements hold due to DI. Applying DIII we can $\mathfrak{A}_0$-derive:

$$\mathfrak{A}_0 \vdash (v_0\colon G \to \text{while}(g(x)) \stackrel{\mathrm{e}}{=} \text{while}(g(g(x)))). \qquad (*)$$

This is not yet the elementary implication we have in mind. For a second application of DIII we fix the parameters as follows:
$(w\colon H \to t_1 \overset{e}{=} t_2)$ the same as above, but $\boldsymbol{t} \in T(\Sigma, v_0)_{w_0}$ is now given by $y\boldsymbol{t} = x$. Evidently, there is an existence equation $(v_0\colon r_1 \overset{e}{=} r_2)$ with $\mathfrak{A}_0 \vdash (v_0\colon G \to r_1 \overset{e}{=} r_2)$ so that x is a subterm of r_1 or r_2. This is obtained by taking $(v_0\colon r_1 \overset{e}{=} r_2) = (v_0\colon p(x) \overset{e}{=} \text{true})$. Condition (c) now requires us to to prove that

$$\mathfrak{A}_0 \vdash (v_0\colon G \to p(x) \overset{e}{=} \text{true})$$

and

$$\mathfrak{A}_0 \vdash (v_0\colon G \to \text{while}(g(x)) \overset{e}{=} \text{while}(g(x))).$$

The first holds due to DI and the second because of (*) above and Corollary 3.1.4. Applying DIII we now $\mathfrak{A}_0$-derive:

$$\mathfrak{A}_0 \vdash (v_0\colon G \to \text{while}(x) \overset{e}{=} \text{while}(g(x))). \qquad (**)$$

Finally, a third application of DIII will produce the intended elementary implication. Now we fix the parameters of DIII as follows:

$$(w\colon H \to t_1 \overset{e}{=} t_2) = (v_M\colon x \overset{e}{=} y,\, y \overset{e}{=} z \to x \overset{e}{=} z) \text{ with } v_M\colon \{x, y, z\} \to M,$$

and $\boldsymbol{t} \in T(\Sigma, v_0)_{v_M}$ is given by

$$x\boldsymbol{t} = \text{while}(x),\; y\boldsymbol{t} = \text{while}(g(x)),\; z\boldsymbol{t} = \text{while}(g(g(x))).$$

Condition (b) is satisfied because of (*), (**), and DI, and also condition (c) is satisfied due to (**) and (*), so that DIII produces

$$\mathfrak{A}_0 \vdash (v_0\colon G \to \text{while}(x) \overset{e}{=} \text{while}(g(g(x)))).$$

The next definition extends the concept of $\mathfrak{A}$-derivability from elementary implications to arbitrary implications. As preparation we consider more thoroughly terms on infinite S-sorted systems of variables. Since every operator $\sigma \in \Omega$ has a finite arity $\sigma\colon w \to s$ with $w\colon [n] \to S$, for every term $t \in T(\Sigma, v)$, where $v\colon x \to S$ is an arbitrary S-sorted system of variables, there exists a finite subsystem $w\colon Y \to S$ of $v\colon X \to S$ with

$$t \in T(\Sigma, w) \subseteq T(\Sigma, v).$$

We say that $w\colon Y \to S$ is a finite subsystem of $v\colon X \to S$ if Y is a finite subset of X and $yw = yv$ for every $y \in Y$.

If $t \in T(\Sigma, v)$ and $w\colon Y \to S$ are related in this way, then we say that w *covers* t, denoted by

$$w \succ t.$$

We say that a finite S-sorted system of variables $w\colon Y \to S$ covers the existence equation $(v\colon t_1 \overset{e}{=} t_2)$ or the set of existence equations $(v\colon G)$, if w

covers as well t_1 as t_2 or if w covers every existence equation of $(v\!:\!G)$, respectively. These facts will be denoted by

$$w \succ (v\colon t_1 \overset{\mathrm{e}}{=} t_2) \quad \text{or} \quad w \succ (v\colon G),$$

respectively. If $w \succ (v\colon t_1 \overset{\mathrm{e}}{=} t_2)$, or $w \succ (v\colon G)$, with $v\colon X \to S$, $w\colon Y \to S$, then we denote by

$$(v{\downarrow}w\colon t_1 \overset{\mathrm{e}}{=} t_2) \quad \text{or} \quad (v{\downarrow}w\colon G)$$

the corresponding existence equation or set of existence equations on the domain $w\colon Y \to S$, which results by forgetting all those variables of X that are not contained in Y. If we use the notation $(v{\downarrow}w\colon t_1 \overset{\mathrm{e}}{=} t_2)$ or $(v{\downarrow}w\colon G)$, then we state simultaneously

$$w \succ (v\colon t_1 \overset{\mathrm{e}}{=} t_2) \quad \text{or} \quad w \succ (v\colon G),$$

respectively.

Definition 3.1.7. Let $\mathfrak{A}$ be any set of elementary implications and $v\colon X \to S$ any S-sorted system of variables. A set of existence equations $(v\colon H)$ is said to be $\mathfrak{A}$-*derivable* from a set of existence equations $(v\colon G)$, denoted by

$$\mathfrak{A} \vdash (v\colon G \to H),$$

if for every $(v\colon t_1 \overset{\mathrm{e}}{=} t_2) \in (v\colon H)$ there exists a finite S-sorted subsystem $w\colon Y \to S$ of $v\colon X \to S$ and a finite subset $(v\colon G^*)$ of $(v\colon G)$ such that

$$w \succ (v\colon G^*), \qquad w \succ (v\colon t_1 \overset{\mathrm{e}}{=} t_2)$$

and

$$\mathfrak{A} \vdash (v{\downarrow}w\colon G^* \to t_1 \overset{\mathrm{e}}{=} t_2). \qquad \blacksquare$$

A useful property of $\mathfrak{A}$-derivability of implications is given by the following corollary.

Corollary 3.1.8. *If $(w\colon F)$, $(w\colon H)$ are sets of existence equations with $w\colon Y \to S$ and $H \neq \emptyset$, if $(v\colon G)$ with $v\colon X \to S$ is a further set of existence equations, and if $t \in T(\Sigma, v)_w$ and $\mathfrak{A}$ any set of elementary implications, then*

$$\mathfrak{A} \vdash (v\colon G \to H\bar{\boldsymbol{t}}) \quad \textit{and} \quad \mathfrak{A} \vdash (w\colon H \to F)$$

imply

$$\mathfrak{A} \vdash (v\colon G \to F\bar{\boldsymbol{t}}).$$

Proof. Let $(w\colon t_1 \overset{\mathrm{e}}{=} t_2)$ be any existence equation of $(w\colon F)$, then $(v\colon t_1\bar{\boldsymbol{t}} \overset{\mathrm{e}}{=} t_2\bar{\boldsymbol{t}})$ is an arbitrary existence equation of $(v\colon F\bar{\boldsymbol{t}})$.

Because $\mathfrak{A} \vdash (w\colon H \to F)$ there exists a finite S-sorted subsystem $u\colon Z \to S$ of $w\colon Y \to S$ and a finite subset $H^* \subseteq H$ with $u \succ (w\colon H^*)$, $u \succ (w\colon t_1 \overset{\mathrm{e}}{=} t_2)$ and $\mathfrak{A} \vdash (w{\downarrow}u\colon H^* \to t_1 \overset{\mathrm{e}}{=} t_2)$. By means of EDIII this implies

$$\mathfrak{A} \vdash (v\colon H^*\bar{\boldsymbol{t}} \to t_1\bar{\boldsymbol{t}} \overset{\mathrm{e}}{=} t_2\bar{\boldsymbol{t}}).$$

Because of $\mathfrak{A} \vdash (v\colon G \to H\tilde{\boldsymbol{t}})$ and of the finiteness of $(w\colon H^*)$ there exists a finite S-sorted subsystem $v^*\colon X^* \to S$ of $v\colon X \to S$ and a finite subset $G^* \subseteq G$ such that

$$v^* \succ (v\colon G^*), \quad v^* \succ (v\colon H^*\tilde{\boldsymbol{t}}), \quad v^* \succ (v\colon t_1\tilde{\boldsymbol{t}} \stackrel{e}{=} t_2\tilde{\boldsymbol{t}})$$

and

$$\mathfrak{A} \vdash (v{\downarrow}v^*\colon G^* \to H^*\tilde{\boldsymbol{t}}).$$

From

$$\mathfrak{A} \vdash (v{\downarrow}v^*\colon G^* \to H^*\tilde{\boldsymbol{t}})$$

and

$$\mathfrak{A} \vdash (v\colon H^*\tilde{\boldsymbol{t}} \to t_1\tilde{\boldsymbol{t}} \stackrel{e}{=} t_2\tilde{\boldsymbol{t}}),$$

$$\mathfrak{A} \vdash (v{\downarrow}v^*\colon G^* \to t_1\tilde{\boldsymbol{t}} = t_2\tilde{\boldsymbol{t}})$$

follows by means of EDIII.

Consequently, for every $(v\colon t_1\tilde{\boldsymbol{t}} \stackrel{e}{=} t_2\tilde{\boldsymbol{t}}) \in (v\colon F\tilde{\boldsymbol{t}})$ there exists a finite S-sorted subsystem $v^*\colon X^* \to S$ of $v\colon X \to S$ and a finite $G^* \subseteq G$ with

$$v^* \succ (v\colon G^*), \qquad v^* \succ (v\colon t_1\tilde{\boldsymbol{t}} \stackrel{e}{=} t_2\tilde{\boldsymbol{t}})$$

and

$$\mathfrak{A} \vdash (v\colon G^* \to t_1\tilde{\boldsymbol{t}} \stackrel{e}{=} t_2\tilde{\boldsymbol{t}}),$$

which implies

$$\mathfrak{A} \vdash (v\colon G \to F\tilde{\boldsymbol{t}}). \qquad \blacksquare$$

The proof of correctness or soundness concludes this section. The completeness of $\mathfrak{A}$-derivability will be proved in the following section.

Theorem 3.1.9. *Let $\mathfrak{A}$ be any set of elementary implications and $(v\colon G \to H)$ with $v\colon X \to S$ any implication. For every $\mathfrak{A}$-algebra A the statement*

$$\mathfrak{A} \vdash (v\colon G \to H) \quad \textit{implies} \quad A_{(v:G)} \subseteq A_{(v:H)}.$$

Proof. The proof is based again on the inductive representation

$$\mathrm{IMP}(\mathfrak{A}) = \bigcup_{i=0}^{\infty} \mathrm{IMP}_i(\mathfrak{A}).$$

First we show that $\mathfrak{A}$-derivability is sound for elementary implications. Let

$$(w\colon G \to t_1 \stackrel{e}{=} t_2) \in \mathrm{IMP}_0(\mathfrak{A}),$$

then $w\colon Y \to S$ is a finite S-sorted system of variables. Therefore either $(w\colon t_1 \stackrel{e}{=} t_2) \in (w\colon G)$ or $t_1 = t_2 = y \in Y$.

Because

$$A_{(w:G)} = \bigcap \{A_{(w:t_1 \stackrel{e}{=} t_2)} \mid (w\colon t_1 \stackrel{e}{=} t_2) \in (w\colon G)\},$$

$$A_{(w:G)} \subseteq A_{(w:t_1 \stackrel{e}{=} t_2)}$$

holds for every $(w: t_1 \overset{e}{=} t_2) \in (w: G)$. Since $(w: G)$ is a set of existence equations on $w: Y \to S$,

$$A_{(w:G)} \subseteq A_w = A_{(w:y \overset{e}{=} y)}$$

holds for every $y \in Y$.

Now we assume that Theorem 3.1.9 is true for every elementary implication of $\mathrm{IMP}_i(\mathfrak{A})$, $i \geq 0$.

If

$$(w: G \to t_1 \overset{e}{=} t_2) \in \mathrm{IMP}_{i+1}(\mathfrak{A}),$$

then there are

$$(u: H \to r_1 \overset{e}{=} r_2) \in \mathrm{TAUT}(\Sigma) \cup \mathfrak{A}, \quad u: Z \to S, \quad \boldsymbol{t} \in T(\Sigma, w)_u$$

such that conditions (b) and (c) of DIII are satisfied with respect to $(w: G)$, and $r_1\tilde{\boldsymbol{t}} = t_1$, $r_2\tilde{\boldsymbol{t}} = t_2$. Consequently, for every $z \in Z$ there is an existence equation $(w: q_1 \overset{e}{=} q_2)$ with

$$(w: G \to q_1 \overset{e}{=} q_2) \in \mathrm{IMP}_i(\mathfrak{A})$$

and $z\boldsymbol{t}$ is a subterm of q_1 or q_2. Additionally

$$(w: G \to p_1\tilde{\boldsymbol{t}} \overset{e}{=} p_2\tilde{\boldsymbol{t}}) \in \mathrm{IMP}_i(\mathfrak{A})$$

holds for every $(u: p_1 \overset{e}{=} p_2) \in (u: H)$. According to the notion of an $\mathfrak{A}$-algebra

$$A_{(u:H)} \subseteq A_{(u:r_1 \overset{e}{=} r_2)}$$

holds for every $(u: H \to r_1 \overset{e}{=} r_2) \in \mathfrak{A}$, and without any problems one can show that

$$A_{(u:H)} \subseteq A_{(u:r_1 \overset{e}{=} r_2)}$$

holds for every tautological elementary implication.

Now $\boldsymbol{a} \in A_{(w:G)}$ may be any solution. Then we have to show that

$$\boldsymbol{a} \in A_{(w:t_1 \overset{e}{=} t_2)}.$$

The existence of an existence equation $(w: q_1 \overset{e}{=} q_2)$ with

$$(w: G \to q_1 \overset{e}{=} q_2) \in \mathrm{IMP}_i(\mathfrak{A})$$

for every $z \in Z$ so that $z\boldsymbol{t}$ is a subterm of q_1 or q_2 implies $z\boldsymbol{t} \in \mathrm{dom}\, \tilde{\boldsymbol{a}}$ for every $z \in Z$. Hence , we can build the assignment $\boldsymbol{t}\tilde{\boldsymbol{a}} \in A_u$. For this assignment and for every $r \in \mathrm{dom}(\widetilde{\boldsymbol{t}\tilde{\boldsymbol{a}}})$ we have

$$\mathrm{dom}(\widetilde{\boldsymbol{t}\tilde{\boldsymbol{a}}}) = (\mathrm{dom}\, \tilde{\boldsymbol{a}})\, \tilde{\boldsymbol{t}}^{-1} \quad \text{and} \quad r(\widetilde{\boldsymbol{t}\tilde{\boldsymbol{a}}}) = (r\tilde{\boldsymbol{t}})\, \tilde{\boldsymbol{a}}.$$

For every $(u: p_1 \overset{e}{=} p_2) \in (u: H)$ we obtain

$$p_1\tilde{\boldsymbol{t}} \in \mathrm{dom}\, \tilde{\boldsymbol{a}}, \quad p_2\tilde{\boldsymbol{t}} \in \mathrm{dom}\, \tilde{\boldsymbol{a}}, \quad \text{and} \quad (p_1\tilde{\boldsymbol{t}})\, \tilde{\boldsymbol{a}} = (p_2\tilde{\boldsymbol{t}})\, \tilde{\boldsymbol{a}}$$

because

$$(w: G \to p_1\bar{\boldsymbol{t}} \overset{e}{=} p_2\bar{\boldsymbol{t}}) \in \mathrm{IMP}_i(\mathfrak{A}).$$

This proves $\boldsymbol{t}\tilde{\boldsymbol{a}} \in A_{(u:H)}$.

$(u: H \to r_1 \overset{e}{=} r_2) \in \mathrm{TAUT}(\Sigma) \cup \mathfrak{A}$ guarantees $\boldsymbol{t}\tilde{\boldsymbol{a}} \in A_{(u:r_1 \overset{e}{=} r_2)}$, and so

$$r_1, r_2 \in \mathrm{dom}(\widetilde{\boldsymbol{t}\tilde{\boldsymbol{a}}}), \qquad r_1(\widetilde{\boldsymbol{t}\tilde{\boldsymbol{a}}}) = r_2(\widetilde{\boldsymbol{t}\tilde{\boldsymbol{a}}}).$$

This together with $t_1 = r_1\bar{\boldsymbol{t}}$, $t_2 = r_2\bar{\boldsymbol{t}}$ yields

$$t_1\tilde{\boldsymbol{a}} = (r_1\bar{\boldsymbol{t}})\,\tilde{\boldsymbol{a}} = r_1(\widetilde{\boldsymbol{t}\tilde{\boldsymbol{a}}}) = r_2(\widetilde{\boldsymbol{t}\tilde{\boldsymbol{a}}}) = (r_2\bar{\boldsymbol{t}})\,\tilde{\boldsymbol{a}} = t_2\tilde{\boldsymbol{a}}$$

or in other words

$$\boldsymbol{a} \in A_{(w:t_1 \overset{e}{=} t_2)}.$$

Finally we extend the proof from elementary implications to arbitrary implications. Therefore let $(v: G \to H)$ with $v: X \to S$ be any implication with $\mathfrak{A} \vdash (v: G \to H)$, and A may be any $\mathfrak{A}$-algebra. We have to prove that

$$\boldsymbol{a} \in A_{(v:t_1 \overset{e}{=} t_2)} \quad \text{results from} \quad \boldsymbol{a} \in A_{(v:G)}$$

for every $(v: t_1 \overset{e}{=} t_2) \in (v: H)$.

Corresponding to Definition 3.1.7 there is a finite S-sorted subsystem $w: Y \to S$ of $v: X \to S$ and a finite subset $G^* \subseteq G$ with

$$w \succ (v: G^*), \qquad w \succ (v: t_1 \overset{e}{=} t_2),$$

and

$$\mathfrak{A} \vdash (v{\downarrow}w: G^* \to t_1 \overset{e}{=} t_2).$$

The restriction of the assignment $\boldsymbol{a} \in A_w$ to the finite S-sorted subsystem $w: Y \to S$ will be denoted by $\boldsymbol{a}{\downarrow}w \in A_w$.

The covering $w \succ (v: t_1 \overset{e}{=} t_2)$ guarantees that

$$\boldsymbol{a}{\downarrow}w \in A_{(v\downarrow w:t_1 \overset{e}{=} t_2)} \quad \text{implies} \quad \boldsymbol{a} \in A_{(v:t_1 \overset{e}{=} t_2)}.$$

Now let $\boldsymbol{a} \in A_{(v:G)}$. Then $\boldsymbol{a} \in A_{(v:G^*)}$ because $G^* \subseteq G$. The covering $w \succ (v: G^*)$ leads to $\boldsymbol{a}{\downarrow}w \in A_{(v\downarrow w:G^*)}$, and the elementary implication $(v{\downarrow}w: G^* \to t_1 \overset{e}{=} t_2)$ with

$$\mathfrak{A} \vdash (v{\downarrow}w: G^* \to t_1 \overset{e}{=} t_2)$$

yields

$$\boldsymbol{a}{\downarrow}w \in A_{(v\downarrow w:G^*)} \subseteq A_{(v\downarrow w:t_1 \overset{e}{=} t_2)}$$

and so

$$\boldsymbol{a} \in A_{(v:t_1 \overset{e}{=} t_2)}. \qquad \blacksquare$$

3.2. Free partial algebras

In this section we present a useful construction, namely, the construction of an $\mathfrak{A}$-algebra $F(\mathfrak{A}, G, v)$ freely generated by a set of existence equations $(v\colon G)$. Using this construction we can prove the completeness of $\mathfrak{A}$-derivability, the existence of free and relatively free $\mathfrak{A}$-algebras, and finally we can give different quotient constructions.

Definition 3.2.1. Let $\Sigma = (S, \alpha\colon \Omega \to S^* \times S)$ be any signature, $\mathfrak{A}$ any set of elementary implications, $v\colon X \to S$ any S-sorted system of variables, and let $(v\colon G)$ be any set of existence equations. A partial algebra $A \in \mathrm{ALG}(\Sigma)$ is referred to as the *$\mathfrak{A}$-algebra freely generated by* $(v\colon G)$ if

(1) A satisfies all axioms of $\mathfrak{A}$, i.e. $A \in \mathrm{ALG}(\mathfrak{A})$;

(2) There exists a solution

$$\boldsymbol{e} \in A_{(v:G)}$$

such that for every $\mathfrak{A}$-algebra B and every solution

$$\boldsymbol{b} \in B_{(v:G)}$$

there exists exactly one homomorphism $b^{\#}\colon A \to B$ with

$$\boldsymbol{b} = \boldsymbol{e}b^{\#}_v . \qquad \blacksquare$$

Evidently, any two $\mathfrak{A}$-algebras A, B freely generated by one and the same set of existence equations $(v\colon G)$ are isomorphic. On account of this uniqueness up to isomorphism we denote by $F(\mathfrak{A}, G, v)$ an $\mathfrak{A}$-algebra freely generated by $(v\colon G)$. Examples can be found at the end of this section.

In a similar way as free total algebras can be constructed from sets of terms we can construct $F(\mathfrak{A}, G, v)$ also from sets of terms.

Let us start with the construction of a partial Σ-algebra $F(\mathfrak{A}, G, v)$: For every $s \in S$ we set

$$F^*(\mathfrak{A}, G, v)_s = \{(t, G) \mid t \in T(\Sigma, v)_s, \quad \mathfrak{A} \vdash (v\colon G \to t \overset{\mathrm{e}}{=} t)\}.$$

The next step is the construction of an S-equivalence

$$\equiv \; = (\equiv_s \mid s \in S)$$

in the S-set $F^*(\mathfrak{A}, G, v)$ by

$$(t_1, G) \equiv_s (t_2, G) \quad \text{iff} \quad \mathfrak{A} \vdash (v\colon G \to t_1 \overset{\mathrm{e}}{=} t_2).$$

It is easy to see that for every $s \in S$ the binary relation $\equiv_s$ in $F^*(\mathfrak{A}, G, v)$ is reflexive, symmetric and transitive.

Definition 3.2.2. We can define

$$F(\mathfrak{A}, G, v) = F^*(\mathfrak{A}, G, v)/\!\equiv_s \quad \text{for every} \quad s \in S, \tag{a}$$

and denote the elements of $F(\mathfrak{A}, G, v)$, i.e. the equivalence classes, by

$$[t, G] = \{(r, G) \mid \mathfrak{A} \vdash (v\colon G \to t \overset{e}{=} r)\}. \tag{b}$$

n-tuples of elements of $F(\mathfrak{A}, G, v)$ will be abbreviated in the following way:

$$[\boldsymbol{t}, G] = ([1\boldsymbol{t}, G], [2\boldsymbol{t}, G], \ldots, [n\boldsymbol{t}, G]) \in F(\mathfrak{A}, G, v)_w \tag{c}$$

where $w\colon [n] \to S$ and $\boldsymbol{t} \in T(\Sigma, v)_w$.

Additionally to the S-set $F(\mathfrak{A}, G, v)$ we define an Ω-indexed family of fundamental operations in the following way: If $\sigma \in \Omega$, $\sigma\colon w \to s$, $w\colon [n] \to S$ then the domain of the fundamental operation

$$\sigma^{F(\mathfrak{A},G,v)}$$

is defined by (3.2.2.d)

$$\operatorname{dom} \sigma^{F(\mathfrak{A},G,v)} = \{[\boldsymbol{t}, G] \mid \boldsymbol{t} \in T(\Sigma, v)_w$$

$$\mathfrak{A} \vdash (v\colon G \to \boldsymbol{t}\sigma^{T(\Sigma,v)} \overset{e}{=} \boldsymbol{t}\sigma^{T(\Sigma,v)})\} \tag{d}$$

And the application of $\sigma^{F(\mathfrak{A},G,v)}$ is defined by

$$[\boldsymbol{t}, G]\, \sigma^{F(\mathfrak{A},G,v)} = [\boldsymbol{t}\sigma^{T(\mathfrak{A},v)}, G]. \tag{e} \quad \blacksquare$$

The independence of the Definition 3.2.2(d) and (e) from the choice of representatives results immediately from the generalized substitution rule (3.1.5).

Before proving that $F(\mathfrak{A}, G, v)$ is an $\mathfrak{A}$-algebra freely generated by $(v\colon G)$ we give another important property of $F(\mathfrak{A}, G, v)$.

Theorem 3.2.3. *Let* $(w\colon H)$ $w\colon Y \to S$ *be any set of existence equations, where* Y *may be infinite. An assignment*

$$\boldsymbol{f} \in F(\mathfrak{A}, G, v)_w, \quad \textit{given by} \quad y\boldsymbol{f} = [y\boldsymbol{t}, G], \quad \boldsymbol{t} \in T(\Sigma, v)_W$$

for every $y \in Y$, *is a solution of* $(w\colon H)$ *in* $F(\mathfrak{A}, G, v)$ *if and only if*

$$\mathfrak{A} \vdash (v\colon G \to H\tilde{\boldsymbol{t}}).$$

Proof. The case $(w\colon H) = (w\colon \emptyset)$ is trivial because

$$\mathfrak{A} \vdash (v\colon G \to \emptyset).$$

Now let $\boldsymbol{f} \in F(\mathfrak{A}, G, v)_{(w:H)}$. Then

$$\boldsymbol{f} \in \operatorname{dom} t_1^{F(\mathfrak{A},G,v)} \cap \operatorname{dom} t_2^{F(\mathfrak{A},G,v)}$$

and

$$\boldsymbol{f}t_1^{F(\mathfrak{A},G,v)} = \boldsymbol{f}t_2^{F(\mathfrak{A},G,v)}$$

hold for every $(w\colon t_1 \overset{e}{=} t_2) \in (w\colon H)$.

In addition one can derive from Definition 3.2.2(d) and (e) by induction the following equalities:

$$\operatorname{dom} t^{F(\mathfrak{A},G,v)} = \{[\boldsymbol{t}, G] \mid \boldsymbol{t} \in T(\Sigma, v)_u, \qquad \mathfrak{A} \vdash (v\colon G \to t\tilde{\boldsymbol{t}} \stackrel{e}{=} t\tilde{\boldsymbol{t}})\},$$
$$[\boldsymbol{t}, G]\, t^{F(\mathfrak{A},G,v)} = [t\tilde{\boldsymbol{t}}, G]$$

for every $t \in T(\Sigma, u)$, $u\colon Z \to S$. Consequently , we obtain

$$\mathfrak{A} \vdash (v\colon G \to t_1\tilde{\boldsymbol{t}} \stackrel{e}{=} t_1\tilde{\boldsymbol{t}}), \qquad \mathfrak{A} \vdash (v\colon G \to t_2\tilde{\boldsymbol{t}} \stackrel{e}{=} t_2\tilde{\boldsymbol{t}}),$$

and

$$[t_1\tilde{\boldsymbol{t}}, G] = [t_2\tilde{\boldsymbol{t}}, G]$$

for $\boldsymbol{f} = [\boldsymbol{t}, G] \in F(\mathfrak{A}, G, v)_{(w:H)}$, and so

$$\mathfrak{A} \vdash (v\colon G \to t_1\tilde{\boldsymbol{t}} \stackrel{e}{=} t_2\tilde{\boldsymbol{t}})$$

for every $(w\colon t_1 \stackrel{e}{=} t_2) \in (w\colon H)$. This is identical with

$$\mathfrak{A} \vdash (v\colon G \to H\tilde{\boldsymbol{t}}).$$

For the proof of the converse direction we assume

$$\mathfrak{A} \vdash (v\colon G \to H\tilde{\boldsymbol{t}})$$

or

$$\mathfrak{A} \vdash (v\colon G \to t_1\tilde{\boldsymbol{t}} \stackrel{e}{=} t_2\tilde{\boldsymbol{t}})$$

for every $(w\colon t_1 \stackrel{e}{=} t_2) \in (w\colon H)$. As a result of this assumption and Corollary 3.1.4.

$$\mathfrak{A} \vdash (v\colon G \to t_1\tilde{\boldsymbol{t}} \stackrel{e}{=} t_1\tilde{\boldsymbol{t}}), \qquad \mathfrak{A} \vdash (v\colon G \to t_2\tilde{\boldsymbol{t}} \stackrel{e}{=} t_2\tilde{\boldsymbol{t}})$$

follow, and so we obtain

$$[\boldsymbol{t}, G] \in \operatorname{dom} t_1^{F(\mathfrak{A},G,v)} \cap \operatorname{dom} t_2^{F(\mathfrak{A},G,v)}$$

and

$$[\boldsymbol{t}, G] \in F(\mathfrak{A}, G, v)_{(w:t_1 \stackrel{e}{=} t_2)}$$

for every $(w\colon t_1 \stackrel{e}{=} t_2) \in (w\colon H)$, i.e.

$$\boldsymbol{f} = [\boldsymbol{t}, G] \in F(\mathfrak{A}, G, v)_{(w:H)}. \qquad \blacksquare$$

Theorem 3.2.4. *For every set of elementary implications* $\mathfrak{A}$ *and every set of existence equations* $(v\colon G)$, $v\colon X \to S$, *the partial algebra* $F(\mathfrak{A}, G, v)$ *is an* $\mathfrak{A}$*-algebra freely generated by* $(v\colon G)$.

Proof. First of all we prove that $F(\mathfrak{A}, G, v)$ is an $\mathfrak{A}$-algebra. Therefore let $(w\colon H \to t_1 \stackrel{e}{=} t_2)$ with $w\colon Y \to S$ be any axiom of $\mathfrak{A}$ and $\boldsymbol{f} \in F(\mathfrak{A}, G, v)_{(w:H)}$. According to Theorem 3.2.3 this means

$$\mathfrak{A} \vdash (v\colon G \to H\tilde{\boldsymbol{t}})$$

where $\boldsymbol{t} \in T(\Sigma, v)_W$ with $\boldsymbol{f} = [\boldsymbol{t}, G]$.

In addition, $\mathfrak{A} \vdash (w\colon H \to t_1 \overset{e}{=} t_2)$ holds as a consequence of Corollary 3.1.3, so that Corollary 3.1.8 yields

$$\mathfrak{A} \vdash (w\colon G \to t_1\tilde{\boldsymbol{t}} \overset{e}{=} t_2\tilde{\boldsymbol{t}}).$$

A further application of Theorem 3.2.3 leads to

$$\boldsymbol{f} = [\boldsymbol{t}, G] \in F(\mathfrak{A}, G, v)_{(w:t_1 \overset{e}{=} t_2)}.$$

Consequently, $F(\mathfrak{A}, G, v)$ satisfies every axiom of $\mathfrak{A}$.

Next we show that the assignment

$$\boldsymbol{e} \in F(\mathfrak{A}, G, v)_v, \quad \text{given by} \quad x\boldsymbol{e} = [x, G] \quad \text{for every} \quad x \in X,$$

is a solution of $(v\colon G)$ in $F(\mathfrak{A}, G, v)$. For $\boldsymbol{i} \in T(\Sigma, v)_v$ with $x\boldsymbol{i} = x$ for every $x \in X$ the following holds

$$\boldsymbol{e} = [\boldsymbol{i}, G] \quad \text{and} \quad G = G\tilde{\boldsymbol{i}}.$$

This implies

$$\mathfrak{A} \vdash (v\colon G \to G)$$

so that Theorem 3.2.3 yields

$$\boldsymbol{e} \in F(\mathfrak{A}, G, v)_{(v:G)}.$$

Finally we assume that A is any $\mathfrak{A}$-algebra and $\boldsymbol{a} \in A_{(v:G)}$ any solution. According to Definition 3.2.2.(b)

$$\mathfrak{A} \vdash (v\colon G \to t \overset{e}{=} t)$$

holds for every $[t, G] \in F(\mathfrak{A}, G, v)$, and by Theorem 3.1.9 we obtain

$$A_{(v:G)} \subseteq A_{(v:t \overset{e}{=} t)}.$$

Hence, for every $\boldsymbol{a} \in A_{(v:G)}$ it follows that $\boldsymbol{a} \in \operatorname{dom} t^A$, and so we can define the homomorphism

$$a^{\#}\colon F(\mathfrak{A}, G, v) \to A$$

by $[t, G]\, a^{\#} = \boldsymbol{a}t^A$ for every $[t, G] \in F(\mathfrak{A}, G, v)$. It remains to show that this really defines a homomorphism with $\boldsymbol{e}a^{\#}_v = \boldsymbol{a}$. For every $x \in X$,

$$x\boldsymbol{e}a^{\#}_v = (x\boldsymbol{e})\, a^{\#}_v = [x, G]\, a^{\#}_v = \boldsymbol{a}x^A = x\boldsymbol{a}.$$

Now let $\sigma \in \Omega$, with $\sigma\colon w \to s$, $w\colon [n] \to S$, be any operator and

$$[\boldsymbol{t}, G] \in \operatorname{dom} \sigma^{F(\mathfrak{A}, G, v)}.$$

On account of Definition 3.2.2.(d) this implies

$$\mathfrak{A} \vdash (v\colon G \to \boldsymbol{t}\sigma^{T(\Sigma, v)} \overset{e}{=} \boldsymbol{t}\sigma^{T(\Sigma, v)}),$$

and therefore

$$([\boldsymbol{t}, G]\, \sigma^{F(\mathfrak{A}, G, v)})\, a_s^{\#} = [\boldsymbol{t}\sigma^{T(\Sigma, v)}, G]\, a_s^{\#} = \boldsymbol{a}(\boldsymbol{t}\sigma^{T(\Sigma, v)})^A = (\boldsymbol{a}(1\boldsymbol{t})^A, \ldots, \boldsymbol{a}(n\boldsymbol{t})^A)\, \sigma^A$$
$$= (\boldsymbol{t}a_w^{\#})\, \sigma^A.$$

This proves that $a^{\#}\colon F(\mathfrak{A}, G, v) \to A$ is a homomorphism with the required property. Since the S-set $X^* = (X_s^* \mid s \in S)$ given by $X_s^* = \{[x, G] \mid xv = s\}$ is a generating S-set of $F(\mathfrak{A}, G, v)$, the assignment $\boldsymbol{a} \in A_v$ can be extended to a homomorphism in at most one way. ■

By means of the preceding results it is easy to prove the *completeness* of $\mathfrak{A}$-derivability.

Theorem 3.2.5. *For any two sets of existence equations* $(v\colon G)$, $(v\colon H)$,

$$\mathfrak{A} \vdash (v\colon G \to H)$$

holds if and only if $A_{(v:G)} \subseteq A_{(v:H)}$ *is true for every* $\mathfrak{A}$*-algebra* A.

Proof. Since Theorem 3.1.9 represents the necessity, we have only to prove the sufficiency.

If $A_{(v:G)} \subseteq A_{(v:H)}$ is true for every $\mathfrak{A}$-algebra, then we obtain for $F(\mathfrak{A}, G, v)$ the relation

$$F(\mathfrak{A}, G, v)_{(v:G)} \subseteq F(\mathfrak{A}, G, v)_{(v:H)}.$$

For $\boldsymbol{i} \in T(\Sigma, v)_v$ with $x\boldsymbol{i} = x$ for every $x \in X$,

$$[\boldsymbol{i}, G] \in F(\mathfrak{A}, G, v)_{(v:H)}$$

holds, if and only if

$$\mathfrak{A} \vdash (v\colon G \to H\tilde{\boldsymbol{i}}).$$

Since $\boldsymbol{i} = \mathrm{Id}_{T(\Sigma, v)}$ and

$$[\boldsymbol{i}, G]\, F(\mathfrak{A}, G, v)_{(v:G)} \subseteq F(\mathfrak{A}, G, v)_{(v:H)}$$

we have proved

$$\mathfrak{A} \vdash (v\colon G \to H).$$ ■

We illustrate the introduced concept of an $\mathfrak{A}$-algebra freely generated by a set of existence equations by applying it to the description of partial algebras. For this reason we recall how laborious the description of two θ_{CAT}-equoids was in Section 2.4. In Section 2.5 we have shown that the class of small categories can be defined by the signature Σ_{CAT} and by the set of elementary implications $\mathfrak{A}^*_{\mathrm{CAT}}$. Now we are a position to define the θ_{CAT}-equoids A, B of Section 2.4 as $\mathfrak{A}^*_{\mathrm{CAT}}$-algebras freely generated by suitable sets of existence equations.

The small category A of Section 2.4 is isomorphic to

$$F(\mathfrak{A}^*_{\mathrm{CAT}}, \emptyset, v)$$

where $v\colon \{x, y\} \to \{\mathrm{Obj}, \mathrm{Mor}\}$ with $xv = yv = \mathrm{Mor}$.

The small category B of Section 2.4 is isomorphic to

$$F(\mathfrak{A}^*_{\mathrm{CAT}}, G_1, v)$$

with $(v\colon G_1) = (v\colon \{\mathrm{target}(x) \stackrel{\mathrm{e}}{=} \mathrm{source}(y)\})$.

The same small category B can be described by

$$F(\mathfrak{A}^*_{\mathrm{CAT}}, G_2, w)$$

where $w\colon \{x, y, z\} \to \{\mathrm{Obj}, \mathrm{Mor}\}$ with $xw = yw = zw = \mathrm{Mor}$ and

$$(w\colon G_2) = (w\colon \{\mathrm{target}(x) \stackrel{\mathrm{e}}{=} \mathrm{source}(y),\ z \stackrel{\mathrm{e}}{=} \mathrm{comp}(x, y)\}).$$

If we consider the set of existence equations

$$(w\colon G_3) = (w\colon \{\mathrm{target}(x) \stackrel{\mathrm{e}}{=} \mathrm{source}(y),\ \mathrm{source}(x) \stackrel{\mathrm{e}}{=} \mathrm{source}(z),\ \mathrm{target}(y) \stackrel{\mathrm{e}}{=} \mathrm{target}(z)\})$$

then $F(\mathfrak{A}^*_{\mathrm{CAT}}, G_3, w)$ can be represented by the following diagram:

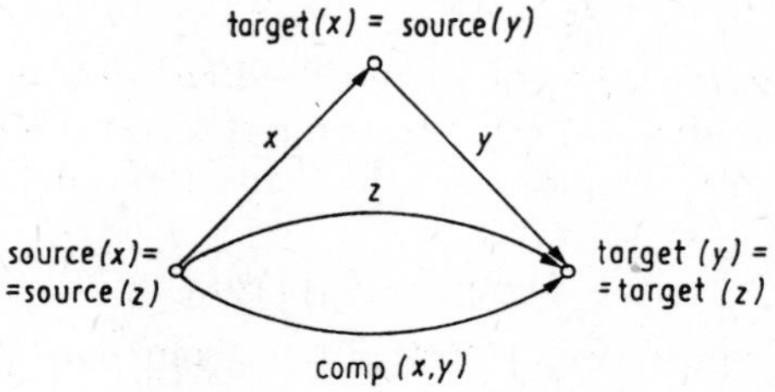

The small category, represented by the following diagram

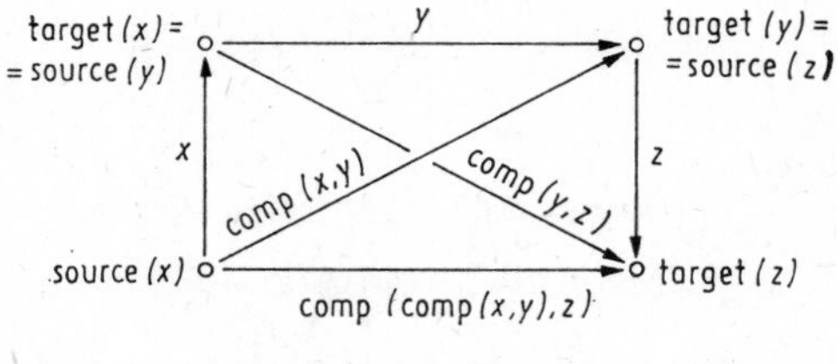

can be described by

$$(w\colon G_4) = (w\colon \{\mathrm{target}(x) \stackrel{\mathrm{e}}{=} \mathrm{source}(y),\ \mathrm{target}(y) \stackrel{\mathrm{e}}{=} \mathrm{source}(z)\}).$$

Finally, $F(\mathfrak{A}^*_{\mathrm{CAT}}, \emptyset, w)$ describes the following small category:

$$\mathrm{source}(x) \xrightarrow{\ x\ } \mathrm{target}(x)$$

$$\mathrm{source}(y) \xrightarrow{\ y\ } \mathrm{target}(y)$$

$$\mathrm{source}(z) \xrightarrow{\ z\ } \mathrm{target}(z)$$

We see that free $\mathfrak{A}^*_{\mathrm{CAT}}$-algebras, i.e. $\mathfrak{A}^*_{\mathrm{CAT}}$-algebras freely generated by an empty set of existence equations $(v\colon \emptyset)$, are of no interest. They coincide with small categories of isolated morphisms.

3.3. Partial quotient algebras

At the end of Section 2.4 we said that the construction of quotient algebras by means of congruence relations cannot be generalized in a straightforward way from total algebras to equoids. We demonstrate this assertion by small categories.

For this reason we consider again the small category A isomorphic to $F(\mathfrak{A}^*_{\mathrm{CAT}}, \emptyset, v)$ with $v\colon \{x, y\} \to \{\mathrm{Obj}, \mathrm{Mor}\}$ and $xv = yv = \mathrm{Mor}$. The two-sorted relation ϱ with

$$\varrho_{\mathrm{Obj}} = \{(e, e) \mid e \in A_{\mathrm{Obj}}\} \cup \{(\mathrm{target}(x), \mathrm{source}(y))\}$$

and

$$\varrho_{\mathrm{Mor}} = \{(m, m) \mid m \in A_{\mathrm{Mor}}\}$$

s a congruence relation, i.e. ϱ is a two-sorted equivalence relation and the carrier of a subequoid of $A \times A$. However, the algebraic structure of A cannot be carried over canonically to the two-sorted quotient set A/ϱ, given by

$$(A/\varrho)_{\mathrm{Mor}} = \{[x]_\varrho, [y]_\varrho\},$$

$$(A/\varrho)_{\mathrm{Obj}} = \{[\mathrm{source}(x)]_\varrho, [\mathrm{target}(x)]_\varrho = [\mathrm{source}(y)]_\varrho, [\mathrm{target}(y)]_\varrho\}$$

Only the unary total operations

$$\mathrm{source}^A\colon A_{\mathrm{Mor}} \to A_{\mathrm{Obj}}, \qquad \mathrm{target}^A\colon A_{\mathrm{Mor}} \to A_{\mathrm{Obj}}$$

can be carried over to A/ϱ by

$$\mathrm{source}^{A/\varrho}([x]_\varrho) = [\mathrm{source}(x)]_\varrho,$$

$$\mathrm{target}^{A/\varrho}([x]_\varrho) = [\mathrm{target}(x)]_\varrho = [\mathrm{source}(y)]_\varrho = \mathrm{source}^{A/\varrho}([y]_\varrho),$$

$$\mathrm{target}^{A/\varrho}([y]_\varrho) = [\mathrm{target}(y)_\varrho].$$

But now there is a non-trivial solution of the domain condition of the binary composition, namely $([x]_\varrho, [y]_\varrho)$, and there is no congruence class that could represent the value of the composition

$$\mathrm{comp}^{A/\varrho}([x]_\varrho, [y]_\varrho).$$

This fact could cause the conjecture that no quotient $\mathfrak{A}^*_{\mathrm{CAT}}$-algebra A/ϱ exists. However, this conjecture is caused only by the lack of success of an elementwise construction. On the other hand the homomorphism $f\colon A \to B$ as defined in Section 2.4 is natural for ϱ.

In the following we show that for every many-sorted relation ϱ, for every set of elementary implications $\mathfrak{A}$, and for every $\mathfrak{A}$-algebra A a homomorphism $f: A \to B$ between $\mathfrak{A}$-algebras exists which is natural for ϱ.

To begin with, we return to the preceding example of the homomorphism $f: A \to B$ natural for ϱ as defined in Section 2.4 on pages 89 and 90. We recall that A is isomorphic to $F(\mathfrak{A}^*_{\mathrm{CAT}}, \emptyset, v)$ and B is isomorphic to

$$F(\mathfrak{A}^*_{\mathrm{CAT}}, \{\mathrm{target}(x) \stackrel{\mathrm{e}}{=} \mathrm{source}(y)\}, v).$$

We can see that up to isomorphism B is constructed from A by adding the ordered pairs of ϱ, considered as existence equations, to the set of existence equations which describes A. In this construction there are no annoing missing elements, because

$$[\mathrm{comp}(x, y), (v: \{\mathrm{target}(x) \stackrel{\mathrm{e}}{=} \mathrm{source}(y)\})]$$

exists in

$$F(\mathfrak{A}^*_{\mathrm{CAT}}, \{\mathrm{target}(x) \stackrel{\mathrm{e}}{=} \mathrm{source}(y)\}, v)$$

due to

$$\mathfrak{A}^*_{\mathrm{CAT}} \vdash (v: \{\mathrm{target}(x) \stackrel{\mathrm{e}}{=} \mathrm{source}(y)\} \to \mathrm{comp}(x, y) \stackrel{\mathrm{e}}{=} \mathrm{comp}(x, y)).$$

In order to investigate the general construction of quotient $\mathfrak{A}$-algebras, we show first that every $\mathfrak{A}$-algebra A is isomorphic to an algebra $F(\mathfrak{A}, G, v)$ where $(v: G)$ can be derived canonically from A.

Let $\mathfrak{A}$ be any set of elementary implications and A any $\mathfrak{A}$-algebra of type $\Sigma = (S, \alpha: \Omega \to S^* \times S)$. Then we define:

$$X_A = \{(a, s) \mid s \in S, a \in A_s\};$$

$$v_A: X_A \to S \quad \text{with} \quad (a, s)\, v_A = s \quad \text{for every} \quad (a, s) \in X_A;$$

$$\boldsymbol{e}_A \in A_{v_A} \quad \text{with} \quad (a, s)\, \boldsymbol{e}_A = a \quad \text{for every} \quad (a, s) \in X_A.$$

Additionally we set

$$G_A = \{(v_A: (a, s) \stackrel{\mathrm{e}}{=} t) \mid t \in T(\Sigma, v_A)_s, s \in S, a = \boldsymbol{e}_A t^A\}.$$

Then $\boldsymbol{e}_A$ is a solution of $(v_A: G_A)$ in A, and the unique extension to a homomorphism

$$e_A^{\#}: F(\mathfrak{A}, G_A, v_A) \to A$$

is an isomorphism:

Corollary 3.3.1.

$$F(\mathfrak{A}, G_A, v_A) \leftrightarrows A.$$

This representation shows that there is no loss of generality if we deal only with $\mathfrak{A}$-algebras of the form $F(\mathfrak{A}, G, v)$. Obviously, there are many different sets of existence equations which describe one and the same

$\mathfrak{A}$-algebra. In general it is recursively undecidable whether $F(\mathfrak{A}, G, v)$ and $F(\mathfrak{A}, H, w)$ are isomorphic, even in the case that $(v: G)$, $(w: H)$ are finite sets of existence equations on finite S-sorted systems of variables.

Now we can start with the construction of a natural homomorphism

$$f: F(\mathfrak{A}, G, v) \to F(\mathfrak{A}, H, w)$$

for every S-sorted relation ϱ in $F(\mathfrak{A}, G, v)$.

Definition 3.3.2. For every binary S-relation ϱ in $F(\mathfrak{A}, G, v)$ we define the so-called *associated set of existence equations* $(v: G_\varrho)$ by

$$(v: G_\varrho) = \{(v: t_1 \stackrel{e}{=} t_2) \mid ([t_1, G], [t_2, G])\, \varrho_s,\ s \in S\}. \qquad \blacksquare$$

Theorem 3.3.3. *For every S-sorted binary relation ϱ in $F(\mathfrak{A}, G, v)$ the homomorphism*

$$f: F(\mathfrak{A}, G, v) \to F(\mathfrak{A}, G \cup G_\varrho, v)$$

with

$$[t, G]\, f = [t, G \cup G_\varrho] \quad \textit{for every} \quad [t, G] \in F(\mathfrak{A}, G, v)$$

is natural for ϱ.

Proof. First we show that the previously defined S-function

$$f: F(\mathfrak{A}, G, v) \to F(\mathfrak{A}, G \cup G_\varrho, v)$$

is a homomorphism. Let $\boldsymbol{e} \in F(\mathfrak{A}, G, v)_{(v:G)}$ be the universal solution of $(v: G)$ i.e. for every $\mathfrak{A}$-algebra B and every solution $\boldsymbol{b} \in B_{(v:G)}$ there exists exactly one homomorphism

$$b^{\#}: F(\mathfrak{A}, G, v) \to B \quad \text{with} \quad \boldsymbol{e}b^{\#}_v = \boldsymbol{b}.$$

The assignment $\boldsymbol{l} \in F(\mathfrak{A}, G \cup G_\varrho, v)_v$ with $x\boldsymbol{l} = [x, G \cup G_\varrho]$ for every $x \in X$ is a solution of $(v: G)$ in $F(\mathfrak{A}, G \cup G_\varrho, v)$ because of $G \subseteq G \cup G_\varrho$ and Theorem 3.2.3. The given S-function

$$f: F(\mathfrak{A}, G, v) \to F(\mathfrak{A}, G \cup G_\varrho, v)$$

is then the unique extension of the solution $\boldsymbol{l} \in F(\mathfrak{A}, G \cup G_\varrho, v)_{(v:G)}$ to a homomorphism. It remains to prove that

$$f: F(\mathfrak{A}, G, v) \to F(\mathfrak{A}, G \cup G_\varrho, v)$$

satisfies condition (2) of Definition 2.4.8.

If B is any $\mathfrak{A}$-algebra and $g: F(\mathfrak{A}, G, v) \to B$ is a homomorphism with $[t_1, G]\, g_s = [t_2, G]\, g_s$ for every $([t_1, G], [t_2, G]) \in \varrho_s$, $s \in S$, then the assignment $\boldsymbol{e}g_v \in B_v$ is a solution of $(v: G \cup G_\varrho)$ in B and $g: F(\mathfrak{A}, G, v) \to B$ is the unique extension of $\boldsymbol{e}g_v \in B_v$ to a homomorphism. Since $\boldsymbol{e}g_v \in B_{(v:G \cup G_\varrho)}$ there exists exactly one homomorphism $h: F(\mathfrak{A}, G \cup G_\varrho, v) \to B$ with

$lh_v = eg_v$. By that

$$f \circ h: F(\mathfrak{A}, G, v) \to B$$

is a homomorphism with

$$e(f \circ h)_v = (ef_v)\, h_v = lh_v = eg_v \quad \text{and} \quad g = f \circ h.$$

If $h': F(\mathfrak{A}, G \cup G_\varrho, v) \to B$ is a further homomorphism with $f \circ h' = g$ then

$$lh'_v = (ef_v)\, h'_v = e(f_v \circ h'_v) = e(f \circ h')_v = eg_v = lh_v$$

and so

$$h = h'. \qquad \blacksquare$$

Theorem 3.3.3 shows that the preceding construction of the natural homomorphism $f: A \to B$ between $\mathfrak{A}^*_{\mathrm{CAT}}$-algebras is representative for constructions of quotient $\mathfrak{A}$-algebras.

Next we are concerned with a second useful construction, the so-called *pushout-construction*, taken from the theory of categories. Let

$$f: A \to B, \qquad g: A \to C$$

be homomorphisms between $\mathfrak{A}$-algebras with common domain A. A couple of homomorphisms

$$f^{\#}: C \to Q, \qquad g^{\#}: B \to Q$$

forms a *pushout of* $f: A \to B$, $g: A \to C$ if

(i) $f \circ g^{\#} = g \circ f^{\#}$;

(ii) for every couple $f': C \to Q'$, $g': B \to Q'$ with $f \circ g' = g \circ f'$ there exists exactly one homomorphism $h: Q \to Q'$ with $g^{\#} \circ h = g'$ and $f^{\#} \circ h = f'$.

Evidently, the homomorphisms $f: A \to B$, $g: A \to C$ determine the pushout-$\mathfrak{A}$-algebra Q uniquely up to isomorphism.

A pictorial representation is the following pushout diagram:

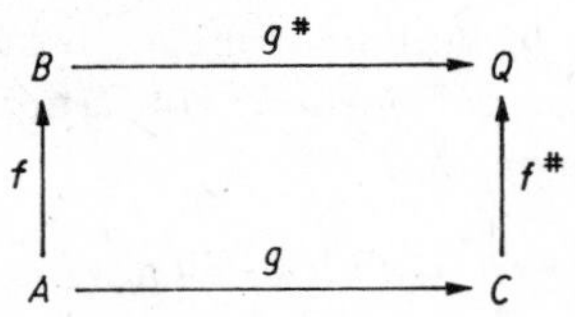

Before giving a general construction of pushout-$\mathfrak{A}$-algebras we show that the pushout generalizes the construction of sums or coproducts of $\mathfrak{A}$-algebras. This construction corresponds to the construction of the free sum of groups in the case of general $\mathfrak{A}$-algebras.

The construction of sums can be derived from the more general pushout construction by means of so-called initial $\mathfrak{A}$-algebras.

Definition 3.3.4. Let $\mathfrak{K} \in \mathrm{ALG}(\mathfrak{A})$ be any class of $\mathfrak{A}$-algebras. An $\mathfrak{A}$-algebra A is called *initial* with respect to the class $\mathfrak{K}$, or *initial* $\mathfrak{K}$-*algebra*, if

(i) $A \in \mathfrak{K}$;

(ii) For every $B \in \mathfrak{K}$ there exists exactly one homomorphism $b: A \to B$. ■

Corollary 3.3.5. *For every set of elementary implications* $\mathfrak{A}$ *the* $\mathfrak{A}$-*algebra* $F(\mathfrak{A}, \emptyset, \lambda)$ *is initial with respect to* $\mathrm{ALG}(\mathfrak{A})$.

Proof. Let B be any $\mathfrak{A}$-algebra. Then there exists exactly one assignment of the empty string $\lambda: \emptyset \to S$ in the carrier of the $\mathfrak{A}$-algebra B, namely the inclusion of the empty set into B. This unique assignment is the only solution of $(\lambda, \emptyset)$ in B, so that by Theorem 3.2.4 exactly one homomorphism $b: F(\mathfrak{A}, \emptyset, \lambda) \to B$ exists. ■

We point out that $F(\mathfrak{A}, \emptyset, \lambda)$ may be the empty algebra or that some of its carrier sets may be empty. For instance,

$$F(\mathfrak{A}^*_{\mathrm{CAT}}, \emptyset, \lambda)$$

is the empty category.

Now let $a: F(\mathfrak{A}, \emptyset, \lambda) \to A$, $b: F(\mathfrak{A}, \emptyset, \lambda) \to B$ be homomorphisms between $\mathfrak{A}$-algebras. According to Corollary 3.3.5 these homomorphisms are uniquely determined by $A, B \in \mathrm{ALG}(\mathfrak{A})$. If $f: A \to C$, $g: B \to C$ are homomorphisms between $\mathfrak{A}$-algebras with common codomain, then in any case the compositions $a \circ f$ and $b \circ g$ are equal and coincide with the unique homomorphism $c: F(\mathfrak{A}, \emptyset, \lambda) \to C$.

An element-free description of the sum $A + B$ is given by homomorphisms $i: A \to A + B$, $j: B \to A + B$ such that for each pair of homomorphisms $f: A \to C$, $g: B \to C$ there exists exactly one homomorphism $h: A + B \to C$ with $i \circ h = f$ and $j \circ h = g$.

If we now consider the pushout of the uniquely determined homomorphisms $a: F(\mathfrak{A}, \emptyset, \lambda) \to A$, $b: F(\mathfrak{A}, \emptyset, \lambda) \to B$ given by the homomorphisms $a^{\#}: B \to Q$, $b^{\#}: A \to Q$, then the $\mathfrak{A}$-algebras Q and $A + B$ are isomorphic. Therefore the construction of the sum of $\mathfrak{A}$-algebras is a special case of the construction of a pushout of homomorphisms between $\mathfrak{A}$-algebras.

Now we describe the general pushout construction. For this purpose we introduce a notational convention. If

$$f: F(\mathfrak{A}, G_1, v_1) \to F(\mathfrak{A}, G_2, v_2)$$

is a homomorphism and $[t, G_1]$ any element of $F(\mathfrak{A}, G_1, v_1)$, then tf denotes any term with $[t, G_1] f = [tf, G_2]$.

Corollary 3.3.6. *Let homomorphisms*

$$f: F(\mathfrak{A}, G_1, v_1) \to F(\mathfrak{A}, G_2, v_2),$$

$$g: F(\mathfrak{A}, G_1, v_1) \to F(\mathfrak{A}, G_3, v_3)$$

be given with

$$v_1 : X \to S, \quad v_2 : Y \to S, \quad v_3 : Z \to S$$

and

$$Y \cap Z = \emptyset.$$

Then

$$f^{\#} : F(\mathfrak{A}, G_3, v_3) \to F(\mathfrak{A}, G, v_2 + v_3)$$

and

$$g^{\#} : F(\mathfrak{A}, G_2, v_2) \to F(\mathfrak{A}, G, v_2 + v_3)$$

form a pushout for f and g, where

$$v_2 + v_3 : Y \cup Z \to S$$

with

$$x(v_2 + v_3) = xv_2 \quad \text{if} \quad x \in Y$$

and

$$x(v_2 + v_3) = xv_3 \quad \text{if} \quad x \in Z,$$

and where

$$G = G_2 \cup G_3 \cup \{(v_2 + v_3 : xf \stackrel{e}{=} xg) \mid x \in X\}.$$

The homomorphisms $f^{\#}$ and $g^{\#}$ are defined by

$$[t, G_2]\, g^{\#} = [t, G] \quad \text{for every} \quad [t, G_2] \in F(\mathfrak{A}, G_2, v_2)$$

and

$$[t, G_3]\, f^{\#} = [t, G] \quad \text{for every} \quad [t, G_3] \in F(\mathfrak{A}, G_3, v_3).$$

Proof. $f^{\#}$ and $g^{\#}$ are homomorphisms according to the same reasoning as in the proof of Theorem 3.3.3. So it remains to show that

$$f \circ g^{\#} = g \circ f^{\#}$$

and that $f^{\#}$ and $g^{\#}$ form a pushout for f and g.

Let $\boldsymbol{l} \in F(\mathfrak{A}, G_1, v_1)_{(v_1 : G_1)}$ denote the universal solution. If we can prove $\boldsymbol{l}(f \circ g^{\#})_{v_1} = \boldsymbol{l}(g \circ f^{\#})_{v_1}$, then $f \circ g^{\#} = g \circ f^{\#}$ follows. According to condition (2) of Definition 3.2.1 $[xf, G] = [xg, G]$ holds for every $x \in X$, since

$$(v_2 + v_3 : xf \stackrel{e}{=} xg) \in (v_2 + v_3 : G) \quad \text{for every} \quad x \in X.$$

This yields

$$\begin{aligned} x\big(\boldsymbol{l}(f \circ g^{\#})\big) &= (x\boldsymbol{l})\,(f \circ g^{\#}) = ([x, G_1]\, f)\, g^{\#} = [xf, G_2]\, g^{\#} \\ &= [xf, G] = [xg, G] = x\big(\boldsymbol{l}(g \circ f^{\#})\big), \end{aligned}$$

and so

$$\boldsymbol{l}(f \circ g^{\#}) = \boldsymbol{l}(g \circ f^{\#}) \quad \text{and} \quad f \circ g^{\#} = g \circ f^{\#}.$$

Now let

$$g' : F(\mathfrak{A}, G_2, v_2) \to B,$$

$$f' : F(\mathfrak{A}, G_3, v_3) \to B$$

be homomorphism with $B \in \mathrm{ALG}(\mathfrak{A})$ and $f \circ g' = g \circ f'$. Then there is exactly one assignment

$$\boldsymbol{b} \in B_{v_2+v_3}$$

with

$$y\boldsymbol{b} = y(\boldsymbol{i}g') \quad \text{for every} \quad y \in Y$$

and

$$z\boldsymbol{b} = z(\boldsymbol{j}f') \quad \text{for every} \quad z \in Z,$$

where

$$\boldsymbol{i} \in F(\mathfrak{A}, G_2, v_2)_{(v:G_2)}, \qquad \boldsymbol{i} \in F(\mathfrak{A}, G_3, v_3)_{(v_3:G_3)}$$

are the universal solutions, respectively.

Because of $G_2 \subseteq G$, $G_3 \subseteq G$, $v_2 \leq v_2 + v_3$, and $v_3 \leq v_2 + v_3$ we can interpret G_2 and G_3 as sets of existence equations on $v_2 + v_3$, and so we obtain

$$\boldsymbol{b} \in B_{(v_2+v_3:G_2)} \quad \text{and} \quad \boldsymbol{b} \in B_{(v_2+v_3:G_3)}.$$

Now let $x \in X$, $t_1 \in T(\Sigma, v_2)$, $t_2 \in T(\Sigma, v_3)$ with $xf = [t_1, G_2]$ and $xg = [t_2, G_3]$, then

$$(v_2 + v_3 \colon t_1 \stackrel{\mathrm{e}}{=} t_2) \in G \setminus (G_2 \cup G_3)$$

because $t_1 = xf$ and $t_2 = xg$. Since $\boldsymbol{b}t_1^B$ does not depend on $z \in Z$, and $\boldsymbol{b}t_2^B$ does not depend on $y \in Y$ the following holds:

$$\begin{aligned} \boldsymbol{b}t_1^B = (\boldsymbol{i}g')\, t_1^B = (\boldsymbol{i}t_1^{F(\mathfrak{A}, G_2, v_2)})\, g' = [t_1, G]\, g' = (xf)\, g' \\ = x(f \circ g') = x(g \circ f') = [t_2, G_3]\, f' = \boldsymbol{b}t_2^B. \end{aligned}$$

This means

$$\boldsymbol{b} \in B_{(v_2+v_3:G)}.$$

Consequently, there is exactly one homomorphism

$$h \colon F(\mathfrak{A}, G, v_2 + v_3) \to B \quad \text{with} \quad \boldsymbol{k}h_{(v_2+v_3)} = \boldsymbol{b},$$

where

$$\boldsymbol{k} \in F(\mathfrak{A}, G, v_2 + v_3)_{(v_2+v_3:G)}$$

denotes the universal solution.

If $\boldsymbol{k} \downarrow v_2$, $\boldsymbol{k} \downarrow v_3$, $\boldsymbol{b} \downarrow v_2$, $\boldsymbol{b} \downarrow v_3$ denote the restrictions of $\boldsymbol{k}$ and $\boldsymbol{b}$ to the S-sorted subsystems v_2 and v_3 of $v_2 + v_3$, respectively, it follows that

$$\boldsymbol{j}f^{\#} = \boldsymbol{k}v_3, \qquad \boldsymbol{i}g^{\#} = \boldsymbol{k}v_2, \qquad \boldsymbol{j}f' = \boldsymbol{b} \downarrow v_3, \qquad \boldsymbol{i}g' = \boldsymbol{b} \downarrow v_2$$

and additionally

$$\begin{aligned} \boldsymbol{j}f^{\#} \circ h = (\boldsymbol{k} \downarrow v_3)\, h = (\boldsymbol{k}h)\, v_3 = \boldsymbol{b} \downarrow v_3 = \boldsymbol{j}g' \\ \boldsymbol{i}g^{\#} \circ h = (\boldsymbol{k} \downarrow v_2)\, h = (\boldsymbol{k}h)\, v_2 = \boldsymbol{b} \downarrow v_2 = \boldsymbol{i}f', \end{aligned}$$

and so

$$f^{\#} \circ h = f' \quad \text{and} \quad g^{\#} \circ h = g'. \qquad \blacksquare$$

Since the pushout construction is not so familiar outside the theory of categories we illustrate this construction by some examples. Furthermore, this construction will be used extensively in Chapter 4 for the construction of complex specifications of abstract data types of simpler ones.

First we remark that the pushout construction encloses the *union of substructures*. Evidently, the set-theoretic union of $\mathfrak{A}$-subalgebras is in general not the carrier of an $\mathfrak{A}$-subalgebra. By the union $A \cup B$ of $\mathfrak{A}$-subalgebras A and B of an $\mathfrak{A}$-algebra C we understand the smallest $\mathfrak{A}$-subalgebra of C containing both A and B. Therefore, the union of $\mathfrak{A}$-subalgebras is the $\mathfrak{A}$-subalgebra generated by the set-theoretic union.

Now let $A \subseteq C$, $B \subseteq C$ be $\mathfrak{A}$-subalgebras of an $\mathfrak{A}$-algebra C, and $i\colon A \cap B \to A$, $j\colon A \cap B \to B$ denote the inclusion of the intersection $A \cap B$ in A and B, respectively. If $i^{\#}\colon B \to Q$ and $j^{\#}\colon A \to Q$ form a pushout of i and j then Q is isomorphic to the union of A and B, or in other words, the union $A \cup B$ with the inclusions of A and B into the structural union $A \cup B$ form a pushout of i and j.

Roughly speaking, the pushout construction of given homomorphisms $f\colon A \to B$, $g\colon A \to C$ corresponds to some kind of structural sticking together of B and C, where the common plane of intersection is determined by $f\colon A \to B$ and $g\colon A \to C$. Surely, $\mathfrak{A}$-algebras are not stuck together, to planes but they are stuck together to $\mathfrak{A}$-algebras.

This interpretation of pushout constructions will be illustrated by means of small categories. Categories are very good for this task, because we can work with finite $\mathfrak{A}$-algebras.

Let the following two-sorted systems of variables be given:

$$v_1\colon \{x\} \to \{\text{Obj}, \text{Mor}\} \quad \text{with} \quad xv_1 = \text{Mor},$$

$$v_2\colon \{y, z\} \to \{\text{Obj}, \text{Mor}\} \quad \text{with} \quad yv_2 = zv_2 = \text{Mor},$$

$$v_3\colon \{y^*, z^*\} \to \{\text{Obj}, \text{Mor}\} \quad \text{with} \quad y^*v_3 = z^*v_3 = \text{Mor}.$$

We start with the homomorphisms

$$f_1\colon F(\mathfrak{A}^*_{\text{CAT}}, \emptyset, v_1) \to F(\mathfrak{A}^*_{\text{CAT}}, \{\text{target}(y) \stackrel{\text{e}}{=} \text{source}(z)\}, v_2)$$

given by $xf_1 = y$;

$$g_1\colon F(\mathfrak{A}^*_{\text{CAT}}, \emptyset, v_1) \to F(\mathfrak{A}^*_{\text{CAT}}, \{\text{target}(y^*) \stackrel{\text{e}}{=} \text{source}(z^*)\}, v_3)$$

given by $xg_1 = y^*$.

The pushout-category of f_1 and g_1 is given by

$$F(\mathfrak{A}^*_{\text{CAT}}, \{y \stackrel{\text{e}}{=} y^*, \text{target}(y) \stackrel{\text{e}}{=} \text{source}(z), \text{target}(y^*) \stackrel{\text{e}}{=} \text{source}(z^*)\}, v_2 + v_3).$$

A graphical representation of this pushout construction is the following:

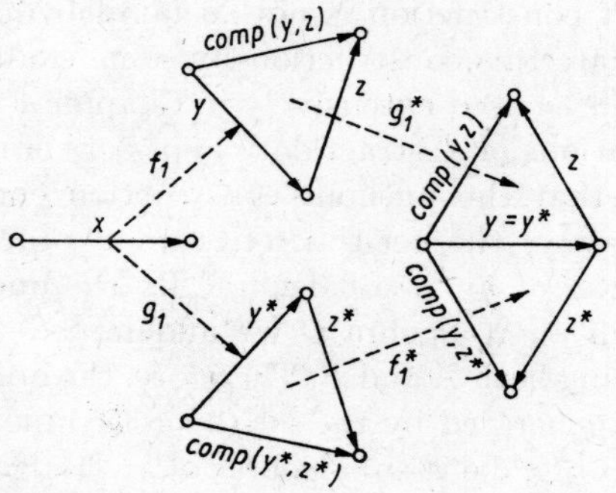

Next we take the homomorphisms

$$f_2: F(\mathfrak{A}^*_{\text{CAT}}, \emptyset, v_1) \to F(\mathfrak{A}^*_{\text{CAT}}, \{\text{target}(y) \overset{\text{e}}{=} \text{source}(z)\}, v_2)$$

given by $xf_2 = y$

$$g_2: F(\mathfrak{A}^*_{\text{CAT}}, \emptyset, v_1) \to F(\mathfrak{A}^*_{\text{CAT}}, \{\text{target}(y^*) \overset{\text{e}}{=} \text{source}(z^*)\}, v_3)$$

given by $xg_2 = z^*$.
Now the pushout-category of f_2 and g_2 is given by

$$F(\mathfrak{A}^*_{\text{CAT}}, \{z^* \overset{\text{e}}{=} y, \text{target}(y) = \text{source}(z), \text{target}(y^*) = \text{source}(z^*)\}, v_2 + v_3)$$

which can be represented graphically as follows:

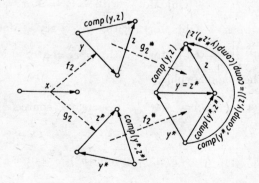

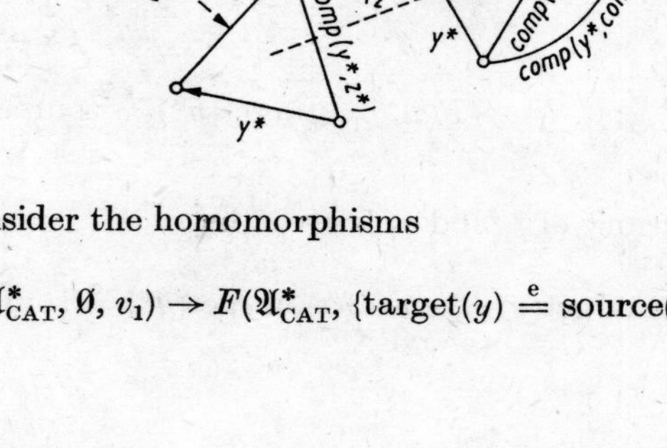

Finally we consider the homomorphisms

$$f_3: F(\mathfrak{A}^*_{\text{CAT}}, \emptyset, v_1) \to F(\mathfrak{A}^*_{\text{CAT}}, \{\text{target}(y) \overset{\text{e}}{=} \text{source}(z)\}, v_2)$$

given by $xf_3 = y$,

$$g_3\colon F(\mathfrak{A}^*_{\mathrm{CAT}}, \emptyset, v_1) \to F(\mathfrak{A}^*_{\mathrm{CAT}}, \{\mathrm{target}(y^*) \stackrel{\mathrm{e}}{=} \mathrm{source}(z^*)\}, v_3)$$

given by $xg_3 = \mathrm{comp}(y^*, z^*)$.
The pushout-category of f_3 and g_3 is given by

$$F(\mathfrak{A}^*_{\mathrm{CAT}}, \{y \stackrel{\mathrm{e}}{=} \mathrm{comp}(y^*, z^*), \mathrm{target}(y) \stackrel{\mathrm{e}}{=} \mathrm{source}(z), \mathrm{target}\ (y^*) \stackrel{\mathrm{e}}{=} \mathrm{source}(z^*)\}, v_2 + v_3).$$

An illustrating diagram is

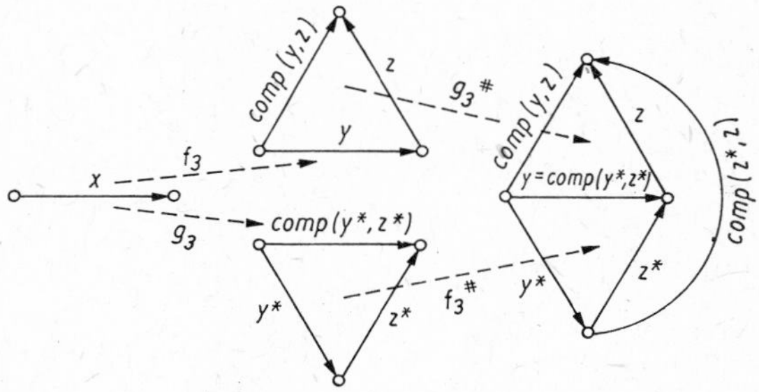

We hope that these examples are helpful for a better understanding of pushout constructions of $\mathfrak{A}$-algebras.

3.4. Theorem of homomorphisms

The quotient construction in Section 3.3 reveals an essential difference in the behaviour of total algebras and of partial ones. A natural homomorphism between $\mathfrak{A}$-algebras is not necessarily a surjective one, i.e. the application of fundamental operations in a quotient $\mathfrak{A}$-algebra cannot be reduced automatically to application of fundamental operations to representatives. Does this fact affect the possibility of the reconstruction of the homomorphic image from the domain and the kernel of the homomorphism?

We have defined the notion of a subalgebra and of the homomorphic image in Section 2.4 only for equoids. Therefore we restrict the considerations of this section also to equoids as distinct from the preceding sections of this chapter.

The results of the preceding section can and will be used in this section. But, we have to take into account the fact that now the set of elementary implications $\mathfrak{A}$ is given as well by the necessary and sufficient domain condition as by axioms interrelating the fundamental operations.

In the case of total algebras the theorem of homomorphisms ensures that exactly the surjective homomorphisms are natural for its kernel.

First we investigate this problem for small categories, i.e. for $\mathfrak{A}_{\mathrm{CAT}}$-equoids or $\mathfrak{A}^*_{\mathrm{CAT}}$-algebras.

Let us consider the surjective homomorphisms

$$f\colon A \to B$$

with

$$A = F(\mathfrak{A}^*_{\mathrm{CAT}}, \emptyset, v),$$

$$B = F(\mathfrak{A}^*_{\mathrm{CAT}}, x_3 \overset{\mathrm{e}}{=} \mathrm{comp}(x_1, x_2), v),$$

$$v\colon \{x_1, x_2, x_3\} \to \{\mathrm{Obj}, \mathrm{Mor}\} \quad \text{with} \quad x_i v = \mathrm{Mor} \quad \text{for} \quad i = 1, 2, 3,$$

$$x_i f_{\mathrm{Mor}} = x_i \quad \text{for} \quad i = 1, 2, 3.$$

The kernel of this homomorphism is given by

$$(\ker f)_{\mathrm{Mor}} = \{(x_i, x_i) \mid i = 1, 2, 3,\}$$

$$(\ker f)_{\mathrm{Obj}} = \{(\mathrm{source}(x_1), \mathrm{source}(x_3)), (\mathrm{target}(x_2), \mathrm{target}(x_3)),$$
$$(\mathrm{target}(x_1), \mathrm{source}(x_2))\} \cup \{(x, x) \mid x \in A_{\mathrm{Obj}}\}.$$

We visualize this situation by the following diagram:

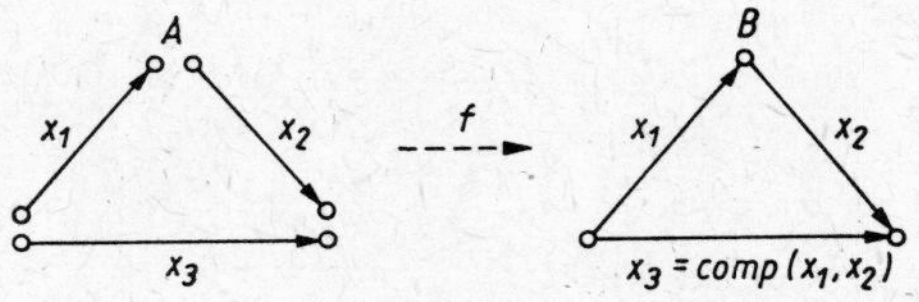

But now $f\colon A \to B$ is not natural for its kernel. For the proof of this assertion we consider the following homomorphism

$$g\colon A \to C$$

with

$$C = F(\mathfrak{A}^*_{\mathrm{CAT}}, G_1, v),$$

$$G_1 = (v\colon \{\mathrm{source}(x_1) \overset{\mathrm{e}}{=} \mathrm{source}(x_3), \mathrm{target}(x_2) \overset{\mathrm{e}}{=} \mathrm{target}(x_3),$$
$$\mathrm{target}(x_1) \overset{\mathrm{e}}{=} \mathrm{source}(x_2)\}),$$

and

$$x_i g = x_i \quad \text{for} \quad i = 1, 2, 3.$$

The small category can be graphically described as follows:

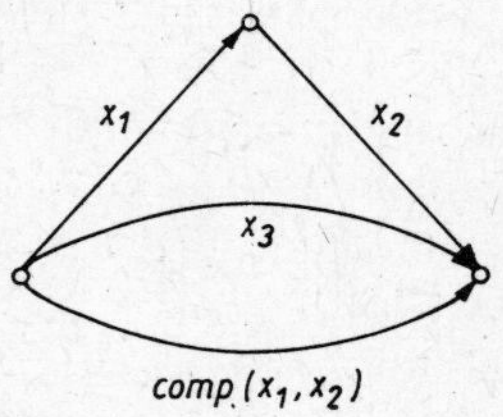

According to Section 3.3 the homomorphism $g\colon A \to C$ is natural for its kernel. Since B and C are not isomorphic and $\ker f = \ker g$, the homomorphism $f\colon A \to B$ cannot be natural for its kernel.

Let $g\colon C \to C'$ now be any homomorphism between small categories, and $q_1\colon C \to Q_1$ may be natural for $\ker g$, so that exactly one homomorphism $h_1\colon Q_1 \to C$ exists with $g = q_1 \circ h_1$. The quotient category Q_1 is isomorphic to the homomorphic image $\langle Cg\rangle$ if and only if $h_1\colon Q_1 \to C$ is injective.

One can generally prove that Q_1 and $\langle Cg\rangle$ are not isomorphic iff somewhere locally in C the situation occures that we described by the homomorphism $f\colon A \to B$. More precisely, $\langle Cg\rangle$ and Q_1 are not isomorphic iff there are morphisms $x, y, z \in C_{\mathrm{Mor}}$ with

$$x \neq \mathrm{id}^C(\mathrm{source}^C(x)), \qquad y \neq \mathrm{id}^C(\mathrm{source}^C(y)),$$

$$z \neq \mathrm{id}^C(\mathrm{source}^C(z)), \qquad \mathrm{target}^C(x) \neq \mathrm{source}^C(y)$$

and

$$zg = \mathrm{comp}^{C'}(xg, yg).$$

In this case zq_1 and $\mathrm{comp}^{Q_1}(xq_1, yq_1)$ are distinct in Q_1, but they have equal images with respect to $h_1\colon Q_1 \to C'$. One can only prove that

$$(q_1)_{\mathrm{Obj}}\colon C_{\mathrm{Obj}} \to (Q_1)_{\mathrm{Obj}}$$

is surjective and

$$(h_1)_{\mathrm{Obj}}\colon (Q_1)_{\mathrm{Obj}} \to C'_{\mathrm{Obj}}$$

is injective. This implies that once more a factorization can be given, but now $h_1\colon Q_1 \to C'$ will be factorized. The result gives the following diagram:

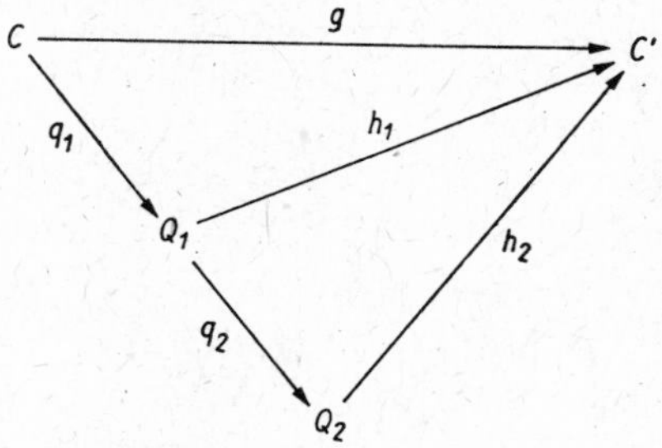

In the second factorization q_2 is natural for $\ker h_1$ and $h_2\colon Q_2 \to C'$ is uniquely determined by $q_2 \circ h_2 = h_1$.

Since $h\colon Q \to C'$ is injective on objects, $h_2\colon Q_2 \to C'$ is injective on morphisms and objects. Therefore the factorization can no longer be continued and Q_2 is isomorphic to $\langle Cg\rangle$. This shows that the homomorphic image of a homomorphism between small categories can in general not be reconstructed by one factorization since two factorizations may be necessary. This existence of iterated quotients motivates the following definition.

Definition 3.4.1. Let θ be any hep-signature and $\mathfrak{A}$ any set of elementary implications. A finite or countable sequence

$$\big((f_i, q_i, f_{i+1}) \mid i \in I\big),$$

of homomorphisms between $\mathfrak{A}$-equoids is called a *chain of iterated quotients of* $f\colon A \to B$, where $A, B \in \mathrm{ALG}(\mathfrak{A})$, if

(1) $f = f_1$ and $f_i = q_i \circ f_{i+1}$ for every $i \in I$;

(2) $q_i\colon Q_i \to Q_{i+1}$ is not an isomorphism for every $i \in I$;

(3) q_i is natural for $\ker f_i$ for every $i \in I$.

If $I = [n]$, then the natural number n is called the *length of the chain of iterated quotients* of $f\colon A \to B$. If alternatively $I = \mathbb{N}$, then we say that the chain of iterated quotients has infinite length. ■

The following diagram may illustrate this definition:

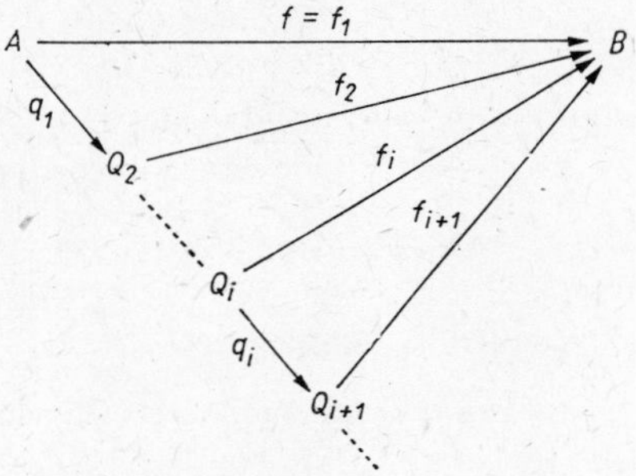

With this notion the *theorem of homomorphism of total algebras* can be formulated as follows: Every chain of iterated quotients of a homomorphism between total algebras is of length one. The preceding considerations show that every chain of iterated quotients of a homomorphism between small categories is at most of length two (see REICHEL (1969)). According to the following theorem the length of chains of iterated quotients is not bounded in the general case.

Theorem 3.4.2. *There are chains of iterated quotients of homomorphisms between equoids of infinite length.*

Proof: Theorem 3.4.2 will be proved by constructing an infinite chain of iterated quotients. According to that let us consider the hep-signature

θ **is sorts** N, P

  oprn $n \to N$

    $s(N) \to N$

    $m(P) \to N$

    $r(x\colon N$ **iff** $xs = n) \to P$

end θ

and let us take the θ-equoids A and B. The θ-equoid A is given by

$$A_N = \{(0, x) \mid x \in \mathbb{N}\}, \qquad A_P = \emptyset,$$

$$n^A = (0, 0),$$

$$(0, x)\, s^A = (0, x+1) \quad \text{for every} \quad x \in \mathbb{N},$$

and

$$\operatorname{dom} m^A = \operatorname{dom} r^A = \emptyset.$$

The θ-equoid B may be defined as follows:

$$B_N = \mathbb{N} \times \mathbb{N}, \qquad B_P = \mathbb{N},$$

$$n^B = (0, 0),$$

$$(x, y)\, s^B = (0, 0) \quad \text{for every} \quad (x, y) \in B_N,$$

$$x m^B = (x+1, 0) \quad \text{for every} \quad x \in B_P,$$

$$\operatorname{dom} r^B = B_N \quad \text{and} \quad (x, y)\, r^B = x \quad \text{for every} \quad (x, y) \in B_N.$$

One can see easily that A is an initial θ-equoid, so that exactly one homomorphism

$$f\colon A \to B$$

exists, which is given by

$$(0, x)\, f_N = (0, 0) \quad \text{for every} \quad (0, x) \in A_N,$$

$$f_P\colon \emptyset \to B_P.$$

Therefore, the kernel of $f\colon A \to B$ is given by

$$(\ker f)_N = A_N \times A_N$$

and

$$(\ker f)_P = A_P \times A_P.$$

By virtue of the construction of natural homomorphisms, according to Theorem 3.3.3 we obtain the first quotient $q_1\colon A \to Q_1$ by setting:

$$(Q_1)_N = \{(0, 0), (1, 0), (1, 1), \ldots, (1, x), \ldots \mid x \in \mathbb{N}\},$$

$$(Q_1)_P = \{0\},$$

$$n^{(Q_1)} = (0, 0),$$

$$(0, 0)\, s^{(Q_1)} = (0, 0),\ (1, x)\, s^{(Q_1)} = (1, x+1) \quad \text{for every} \quad x \in \mathbb{N},$$

$$0\, m^{(Q_1)} = (1, 0),$$

$$\operatorname{dom} r^{(Q_1)} = \{(0, 0)\} \quad \text{and} \quad (0, 0)\, r^{(Q_1)} = 0.$$

The uniquely determined homomorphism

$$f_2\colon Q_1 \to B \quad \text{with} \quad f = f_1 = q_1 \circ f_2$$

is given by

$$(j, 0)\,(f_2)_N = (j, 0) \quad \text{for} \quad j = 0, 1,$$

$$(1, x)\,(f_2)_N = (0, 0) \quad \text{for every} \quad x \in \mathbb{N} \quad \text{with} \quad x \geq 1,$$

and

$$0(f_2)_P = 0.$$

Inductively we define (f_k, q_k, f_{k+1}) for $k \geq 2$ by

$$(A_k)_N = \{(0, 0), (1, 0), \ldots, (k, 0), (k, 1), \ldots, (k, x), \ldots \mid x \in \mathbb{N}\},$$

$$(A_k)_P = \{0, 1, \ldots, k-1\},$$

$$n^{(A_k)} = (0, 0),$$

$$(j, x)\, s^{(A_k)} = \begin{cases} (0, 0) & \text{if} \quad x = 0 \quad \text{and} \quad j \leq k - 1, \\ (k, x+1) & \text{if} \quad k = j, \quad x \in \mathbb{N}, \end{cases}$$

$$j m^{(A_k)} = (j + 1, 0) \quad \text{for} \quad j = 0, 1, \ldots, k - 1,$$

$$\operatorname{dom} r^{(A_k)} = \{(0, 0), (1, 0), \ldots, (k - 1, 0)\},$$

$$(j, 0)\, r^{(A_k)} = j \quad \text{for} \quad j = 0, 1, \ldots, k - 1.$$

The uniquely determined homomorphism

$$f_k\colon Q_k \to B$$

is given by

$$(j, 0)\,(f_k)_N = (j, 0) \quad \text{for} \quad j = 0, 1, \ldots, k,$$

$$(k, x)\,(f_k)_N = (0, 0) \quad \text{for every} \quad x \in \mathbb{N} \quad \text{with} \quad x \neq 0,$$

$$j(f_k)_P = j \quad \text{for} \quad j = 0, 1, \ldots, k - 1,$$

and the homomorphism

$$q_k\colon Q_k \to Q_{k+1}$$

which is natural for $\ker f_k$ is defined by

$$(j, 0)\,(q_k)_N = (j, 0) \quad \text{for} \quad j = 0, 1, \ldots, k,$$

$$(k, x)\,(q_k)_N = (0, 0) \quad \text{for every} \quad x \in \mathbb{N} \quad \text{with} \quad x \neq 0,$$

$$j(q_k)_P = j \quad \text{for} \quad j = 0, 1, \ldots, k - 1.$$

This explicitly defined infinite chain of iterated quotients of a homomorphism $f\colon A \to B$ proves Theorem 3.4.2. ■

Now we are looking at significant relations between the maximal chain of iterated quotients of a homomorphism $f\colon A \to B$ and the homomorphic image $\langle Af \rangle$.

For this reason we consider the homomorphic image $\langle Af \rangle$ of the homomorphism $f\colon A \to B$ of Proof 3.4.2. We see that

$$\langle Af \rangle_N = \{(x, 0) \mid x \in \mathbb{N}\}, \qquad \langle Af \rangle_P = \mathbb{N},$$

so that $\langle Af \rangle$ is an infinite subequoid, whereas the set-theoretic image consists of one element, namely $(0, 0)$. Furthermore we see that with increasing index k the k-th iterated quotient Q_k becomes more and more similar to the homomorphic image. Hence, the following question arises. Can the homomorphic image $\langle Af \rangle$ of a homomorphism be considered as the limit of the maximal chain of iterated quotients of $f: A \to B$?

In both cases of total algebras and small categories, where the maximal chains of iterated quotients are of finite length, this question can be answered positively, because we can evidently define that the last quotient of a finite chain of iterated quotients is the limit of this chain.

But what is the limit of an infinite chain of iterated quotients? The following definition of the limit of a countable chain of morphisms

$$A_1 \xrightarrow{f_1} A_2 \xrightarrow{f_2} A_3 \xrightarrow{f_3} \dots \xrightarrow{f_{n-1}} A_n \xrightarrow{f_n} \dots$$

is derived from the theory of categories, where it corresponds to the colimit of the chain.

Definition 3.4.3. Let θ be a hep-signature, $\mathfrak{A}$ a set of elementary implications, and

$$(f_i: A_i \to A_{i+1} \mid i \in \mathbb{N})$$

a countable chain of $\mathfrak{A}$-equoids. An $\mathfrak{A}$-equoid L, is called a *limit* of this chain, if

(1) there are canonical morphisms

$q_i: A_i \to L$ with $q_i = f_i \circ q_{i+1}$ for every $i \in \mathbb{N}$,

(2) for every family of homomorphisms

$(g_i: A_i \to B)$ with $g_i = f_i \circ g_{i+1}$ for every $i \in \mathbb{N}$

there is exactly one homomorphism

$h: L \to B$ with $g_i = q_i \circ h$ for every $i \in \mathbb{N}$. ■

The situation of Definition 3.4.3 can be illustrated as follows:

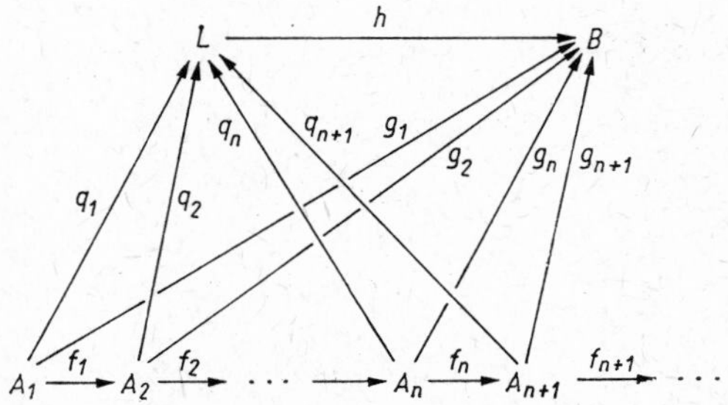

Intuitively, condition (1) requires that L is an upper bound and condition (2) implies that L is a least upper bound of the chain. Additionally, condition (2) guarantees that a limit of a chain is unique up to isomorphism.

If we allow in Definition 3.4.3 finite chains

$$(f_i\colon A_i \to A_{i+1} \mid i \in [n])$$

and modify conditions (1) and (2) for finite index sets, then every finite chain

$$A_1 \xrightarrow{f_1} A_2 \xrightarrow{f_2} \dots \xrightarrow{f_n} A_{n+1}$$

has a limit, namely $L = A_{n+1}$ where the canonical homomorphisms

$$q_i\colon A_i \to L \quad \text{for} \quad i \in [n+1]$$

are defined by

$$q_{n+1} = \mathrm{Id}_{A_{n+1}}, \qquad q_n = f_n, \qquad q_{n-1} = f_{n-1} \circ f_n,$$

$$q_i = f_i \circ f_{i+1} \circ \cdots \circ f_{n-1} \circ f_n, \quad \text{i.e.}$$

$$q_i = f_i \circ q_{i+1} \quad \text{for} \quad i \in [n].$$

Next we are concerned with the existence of limits of infinite chains.

Corollary 3.4.4. *Let θ be a hep-signature and $\mathfrak{A}$ any set of elementary implications. For every infinite chain*

$$(f_i\colon A_i \to A_{i+1} \mid i \in \mathbb{N})$$

of homomorphisms between $\mathfrak{A}$-equoids there exists a limit $L \in \mathrm{ALG}(\theta, \mathfrak{A})$ which is unique up to isomorphism.

Proof. According to general theorems on colimits in categories (see MacLane (1971)) Corollary 3.4.4 would be an immediate consequence of Theorem 3.3.3 and Corollary 3.3.6. But we will give here an explicit construction of L.

If S is the set of sort names of θ, we define for every $s \in S$ the set

$$L_s^* = \{(x, j) \mid j \in \mathbb{N}, x \in (A_j)_s\}.$$

For every ordered pair $(i, j) \in \mathbb{N} \times \mathbb{N}$ with $i < j$ we define a homomorphism

$$f_{i,j}\colon A_i \to A_j \quad \text{by} \quad f_{i,j} = f_i \circ f_{i+1} \circ \cdots \circ f_{j-1}.$$

Now we define an S-equivalence $(\equiv_s \mid s \in S)$ on

$$L^* = (L_s^* \mid s \in S)$$

by

$$(x, i) \equiv_s (y, j) \quad \text{if there is a} \quad k \in \mathbb{N} \quad \text{with} \quad i \leq k, \quad j \leq k$$

and

$$x f_{i,k} = y f_{j,k}.$$

Evidently, this relation is an S-equivalence and so we can construct the quotient sets $L_s = L_s^*/\equiv_s$, where the equivalence classes are denoted by

$$[x, j] = \{(y, i) \mid (x, j) \equiv_s (y, i)\}.$$

In the second step we define the family of fundamental operations

$$(\sigma^L, \sigma \in \Omega)$$

on the S-set $L = (L_s \mid s \in S)$ in the following way: Let $\sigma \in \Omega$, $\sigma\colon w \to s$, $w\colon [n] \to S$. Then we set

$$\operatorname{dom} \sigma^L = \{([x_1, i_1], \ldots, [x_n, i_n])$$

there is a $k \in \mathbb{N}$ and there are

$$y_1, \ldots, y_n \quad \text{with} \quad i_1 \leq k, \ldots, i_n \leq k,$$

$$[x_1, i_1] = [y_1, k], \ldots, [x_n, i_n] = [y_n, k]$$

and

$$(y_1, \ldots, y_n) \in \operatorname{dom} \sigma_{(A_k)}\},$$

and

$$([x_1, i_1], \ldots, [x_n, i_n])\, \sigma^L = [(y_1, \ldots, y_n)\, \sigma^{(A_k)}, k]$$

for every $([x_1, i_1], \ldots, [x_n, i_n]) \in \operatorname{dom} \sigma^L$.

Straightforward considerations show that $\operatorname{dom} \sigma^L$ and the application of σ^L is defined independently of the choice of representatives, and that L is really an $\mathfrak{A}$-equoid.

The canonical morphisms $q_i\colon A_i \to L$ are given by $x(q_i)_s = [x, i]$ for every $s \in S$, $x \in (A_i)_s$, $i \in \mathbb{N}$. If $(g_i\colon A_i \to B \mid i \in \mathbb{N})$ is any other family of homomorphisms with $g_i = f_i \circ g_{i+1}$ for every $i \in \mathbb{N}$, we can define an S-function

$$h\colon L \to B \quad \text{by} \quad [x, i]\, h_s = x(g_i)_s$$

for every $s \in S$, $[x, i] \in L_s$. We show that $h\colon L \to B$ is defined independently of the choice of representatives. If $[x, i] = [y, j] \in L_s$, then there exists a natural number $k \in \mathbb{N}$ with $i \leq k$, $j \leq k$ and $xf_{i,k} = yf_{j,k}$ and so

$$[x, i]\, h = xg_i = x(f_{i,k} \circ g_k) = (xf_{i,k})\, g_k = (yf_{i,k})\, g_k = y(f_{j,k} \circ g_k)$$
$$= yg_j = [y, j]\, h.$$

It remains to show that $h\colon L \to B$ is a homomorphism. Let $\sigma \in \Omega$, $\sigma\colon w \to s$, $w\colon [n] \to S$, and $([x_1, i_1], \ldots, [x_n, i_n])$ in $\operatorname{dom} \sigma^L$. According to the definition of $\operatorname{dom} \sigma^L$ we obtain

$$\big(([x_1, i_1], \ldots, [x_n, i_n])\, \sigma^L\big)\, h = [(y_1, \ldots, y_n)\, \sigma^{(A_k)}, k]\, h = \big((y_1, \ldots, y_n)\, \sigma^{(A_k)}\big)\, g_k$$
$$= (y_1 g_k, \ldots, y_n g_k)\, \sigma^B = ([y_1, k]\, h, \ldots, [y_n, k]\, h)\, \sigma^B$$
$$= ([x_1, i_1]\, h, \ldots, [x_n, i_n]\, h)\, \sigma^B.$$

Additionally $x(q_i \circ h) = [x, i]\, h = xg_i$ holds for every $i \in \mathbb{N}$, i.e. $q_i \circ h = g_i$ holds for every $i \in \mathbb{N}$. If, on the other hand, $h^*\colon L \to B$ is a homomorphism with $q_i \circ h^* = g_i$ for every $i \in \mathbb{N}$, then

$$[x, i]\, h = xg_i = x(q_i \circ h^*) = [x, i]\, h^*$$

for every $i \in \mathbb{N}$ so that finally $h = h^*$. ∎

We want to remark that the θ-equoid L is constructed by the set-theoretic limit of the chain of S-functions on which in a canonical way the additional algebraic structure, namely the Ω-indexed family of fundamental operations, is carried over from the elements of the chain. This leads to the fact that the application of any fundamental operation σ^L to any n-tuple of arguments can be reduced to a suitable element of the chain. Locally we can compute in a suitable element of the chain instead of in the limit of the chain. But we cannot say that different results obtained by local computing in an element of the chain represent different results in the limit structure L. The relations between the elements of a chain and the limit structure can very well be studied by the example of Proof 3.4.2. According to the following theorem of homomorphisms the homomorphic image $\langle Af \rangle$ is always the limit of the chain of iterated quotients.

Theorem 3.4.5 (*Theorem of Homomorphisms*). *Let θ be any hep-signature, $\mathfrak{A}$ any set of elementary implications, and $f\colon A \to B$ any homomorphism between $\mathfrak{A}$-equoids. The homomorphic image $\langle Af \rangle \subseteq B$ is a limit of every maximal chain of iterated quotients of $f\colon A \to B$.*

Proof. If both

$$\big((f_i, q_i, f_{i+1}) \mid i \in [n]\big) \quad \text{and} \quad \big((f_j^*, q_j^*, f_{j+1}^*) \mid j \in [k]\big)$$

are maximal chains of iterated quotients of $f\colon A \to B$, then $n = k$ and there exists a sequence $(h_i\colon Q_i \to Q_i^* \mid i \in [n+1])$ of isomorphisms with

$$h_1 = \mathrm{Id}_A, \quad q_i^* \circ h_{i+1} = h_i \circ q_i \quad \text{for} \quad i = 2, 3, \ldots, n,$$

and

$$f_i^* = h_i \circ f_i \quad \text{for} \quad \text{every} \quad i \in [n+1].$$

This can easily be seen for $i = 2$ and can be proved by induction for all $i \in [n+1]$.

If $\big((f_i, q_i, f_{i+1}) \mid i \in \mathbb{N}\big)$ is a maximal chain of iterated quotients of $f\colon A \to B$, then any other maximal chain of iterated quotients of $f\colon A \to B$ is infinite, too, and is elementwise isomorphic to the given one, in the same way as described above. Because limits of isomorphic chains are also isomorphic, we can start with any maximal chain of iterated quotients of $f\colon A \to B$, which may be given by

$$\big((f_i, q_i, f_{i+1}) \mid i \in J\big).$$

Furthermore, let L be a limit of the chain $(q_i\colon Q_i \to Q_{i+1} \mid i \in J)$ and $(h_i\colon Q_i \to L \mid i \in J)$ may denote the family of canonical homomorphisms. We can assume that L is constructed as the limit in Proof 3.4.4. Without loss of generality we can suppose $\langle Af \rangle = B$. Then there exists a homomorphism

$$h: L \to B \quad \text{with} \quad f_i = h_i \circ h \quad \text{for every} \quad i \in J.$$

According to Theorem 2.4.2 it is sufficient to show that $h: L \to B$ is injective and surjective.

Let $[x, i_1], [y, i_1] \in L_s$, $s \in S$ with $[x, i_1]\, h_s = [y, i_2]\, h_s$. Because $f_i = h_i \circ h$ for every $i \in J$ and $xh_i = [x, i]$, we obtain

$$(xh_{i_1})\, h = (yh_{i_2})\, h, \quad \text{i.e.} \quad xf_{i_1} = yf_{i_2}.$$

If $i_1 < i_2$ we set

$$q: q_{i_1} \circ q_{i_1+1} \circ \cdots \circ q_{i_2-1}: Q_{i_1} \to Q_{i_2}$$

and obtain $xqf_{i_2} = xf_{i_1} = yf_{i_2}$ so that $(x, y) \in \ker f_{i_2}$. Since q_{i_2} is natural for $\ker f_{i_2}$ it follows $xq_{i_2} = yq_{i_2}$. Hence, by definition of L we obtain $(x, i_1) \equiv_s (y, i_2)$, i.e. $[x, i_1] = [y, i_2]$ and so $h: L \to B$ is injective.

Now let $b \in B_s$, $s \in S$, be any element of the homomorphic image. Because $B = \langle Af \rangle$, Corollary 2.4.7 yields the existence of

$$w: [n] \to S, \quad t \in T(\Sigma, w)_s, \quad \boldsymbol{a} \in A_w \quad \text{with} \quad \boldsymbol{a}f_w \in \operatorname{dom} t^B$$

and $b = (\boldsymbol{a}f_w)\, t^B$.

Because each injective homomorphism between equoids reflects solutions of domain conditions, see Theorem 2.4.2, and

$$\big(\boldsymbol{a}(h_1)_w\big)\, h_w = \boldsymbol{a}f_w \in \operatorname{dom} t^B$$

it follows that

$$\Big(\big(\boldsymbol{a}(h_1)_w\big)\, t^L\Big)\, h = \Big(\big(\boldsymbol{a}(h_1)_w\big)\, h_w\Big)\, t^B = (\boldsymbol{a}f_w)\, t^B = b.$$

This proves that $h: L \to B$ is surjective. ■

If we try to prove the theorem of homomorphisms without the hierarchy condition we fail. The bijective homomorphism

$$e: F^* \to F$$

of Proof 2.4.2 can be used for showing that the theorem of homomorphisms is not true in the non-hierarchic case. However, this assertion requires the definition of the notion of homomorphic image for the non-hierarchic case. If we define the homomorphic image $\langle Af \rangle$ to be the subalgebra of B generated by the set-theoretic image Af, recall that then $\langle Af \rangle$ is closed with respect to all fundamental operations of B, then we obtain for the special homomorphism $e: F^* \to F$ the equality $F^*e = F$. For being injective the identity $\mathrm{Id}_{F^*}: F^* \to F^*$ is natural for $\ker e$, so that F^* is the limit of the chain of iterated quotients of $e: F^* \to F$. Since F and F^* are not isomorphic the theorem of homomorphisms is not true.

Because maximal chains of iterated quotients of a homomorphism

$$f: A \to B$$

are of equal length,we define the *homomorphic number* hom f for $f: A \to B$ to be the length of a maximal of iterated quotients. If this chain is infinite we set hom $f = \infty$. For each hep-signature θ and each set of elementary implications we define hom $(\theta, \mathfrak{A})$ to be the supremum of all hom f of homomorphisms $f: A \to B$ between $\mathfrak{A}$-equoids.

It seems to be of interest to give a characterization of those $(\theta, \mathfrak{A})$ that have a finite homomorphic number. However, we do not know any condition which is necessary or sufficient for the finiteness of hom $(\theta, \mathfrak{A})$. We guess that it is recursively undecidable whether hom $(\theta, \mathfrak{A})$ is finite or not.

3.5. Relatively free partial algebras

The aim of this section is the development of a suitable algebraic basis for parameterized abstract data types with partial operations. Examples of parameterized abstract data types are given in Chapter 1. There the concept of an abstract data type is intuitively based on the understanding of a data type as a conceptual unit of a set, the set of admissible values, and a family of fundamental operations on the set of admissible values, where the fundamental operations or combinations of them are the only means for dealing with admissible values. Thus a data type can very well be understood as an algebraic structure. According to that understanding, parameterized data types can be represented by constructions of algebras from given ones, where the base algebras and the constructed algebras have different types in general.

In the first half of this section we introduce appropriate notions for dealing with algebras of different types. Assume

$$\varphi = (\varphi_S, \varphi_\Omega): (S_1, \alpha_1: \Omega_1 \to S_1^* \times S_1) \to (S_2, \alpha_2: \Omega_2 \to S_2^* \times S_2)$$

is a signature morphism (see Definition 2.2.2), $v: X \to S_1$ an S_1-sorted system of variables, and

$$t \in T(\Sigma_1, v)_s \quad \text{with} \quad s \in S_1.$$

Then there is induced an S_2-sorted system of variables

$$v\varphi_S: X \to S_2$$

resulting from the composition of the functions $v: X \to S_1$, $\varphi_S: S_1 \to S_2$. In the following we denote this induced S_2-sorted system of variables briefly by

$$v\varphi: X \to S_2.$$

Definition 3.5.1. There is given an S_1-indexed family of functions

$$\varphi_s\colon T(\Sigma_1, v)_s \to T(\Sigma_2, v\varphi)_{s\varphi_S}, \quad s \in S_1,$$

which are defined inductively as follows:

(a) If $x \in X$ then

$$x\varphi_{xv} = x \in T(\Sigma_2, v\varphi)_{(xv)\varphi_S};$$

(b) If $\sigma \in \Omega_1$, $\sigma\colon \lambda \to s$, then

$$\sigma\varphi_s = \sigma\varphi_\Omega \in T(\Sigma_2, v\varphi)_{s\varphi_S};$$

(c) If $\sigma \in \Omega_1$, $\sigma\colon w \to s$, $\boldsymbol{t} \in T(\Sigma_1, v)_w$ and $t = \boldsymbol{t}\sigma^{T(\Sigma_1,v)}$, then

$$t\varphi_s = (\boldsymbol{t}\varphi_w)\,(\sigma\varphi_\Omega)^{T(\Sigma_2,v\varphi)}. \qquad \blacksquare$$

In the following we allow the omission of subscripts for the functions defined in Definition 3.5.1.

We extend these functions from terms to existence equations, sets of existence equations, and elementary implications by setting

$$(v\colon t_1 \overset{\mathrm{e}}{=} t_2)\,\varphi = (v\varphi\colon t_1\varphi \overset{\mathrm{e}}{=} t_2\varphi),$$

$$(v\colon \emptyset)\,\varphi = (v\varphi\colon \emptyset),$$

$$(v\colon G)\,\varphi = \{(v\varphi\colon t_1\varphi \overset{\mathrm{e}}{=} t_2\varphi) \mid (v\colon t_1 \overset{\mathrm{e}}{=} t_2) \in G\},$$

$$(v\colon G \to t_1 \overset{\mathrm{e}}{=} t_2)\,\varphi = (v\varphi\colon G\varphi \to t_1\varphi \overset{\mathrm{e}}{=} t_2\varphi).$$

Next we extend the notion of a signature morphism to hep-theories. For any hep-theory T we denote by

$$S(T),\ \Omega(T),\ \Theta(T),\ \mathfrak{A}(T),\ \Sigma(T)$$

the set of sort names of T, the set of operators of T, the hep-signature of T, the stable set of elementary implications of T, and the signature of T, respectively.

Definition 3.5.2. A *theory-morphism*

$$\varphi\colon T_1 \to T_2$$

is given by a signature morphism $\varphi\colon \Sigma(T_1) \to \Sigma(T_2)$ so that

(1) $\mathfrak{A}(T_2) \vdash \big(w\varphi\colon \mathrm{def}\,(\sigma\varphi) \to (\mathrm{def}\ \sigma)\,\varphi\big)$ and $\mathfrak{A}(T_2) \vdash \big(w\varphi\colon (\mathrm{def}\,\sigma)\,\varphi \to \mathrm{def}\,(\sigma\varphi)\big)$ for every $\sigma \in \Omega(T_1)$, $\sigma\colon w \to s$,

(2) $\mathfrak{A}(T_2) \vdash (v\colon G \to t_1 \overset{\mathrm{e}}{=} t_2)\,\varphi$ for every $(v\colon G \to t_1 \overset{\mathrm{e}}{=} t_2) \in \mathfrak{A}(T_1)$. $\blacksquare$

For a theory-morphism $\varphi\colon T_1 \to T_2$ it is not required that def $(\sigma\varphi)$ is equal to the image of def σ by φ, but it is required that def $(\sigma\varphi)$ and (def σ) φ are semantically equivalent with respect to $\mathfrak{A}(T_2)$. Analogously we do not suppose that φ maps axioms of T_1 to axioms of T_2, but we require that φ maps an axiom of T_1 to an elementary implication derivable from $\mathfrak{A}(T_2)$.

Next we demonstrate how a T_1-equoid $A\varphi$ from a T_2-equoid A can be derived for a given theory-morphism $\varphi: T_1 \to T_2$. Roughly speaking $A\varphi$ is derived by taking every sort name $s \in S(T_1)$ the $s\varphi_S$-carrier of A and for every operator $\sigma \in \Omega(T_1)$ the fundamental operation $(\sigma\varphi_\Omega)^A$.

Definition 3.5.3. Let $\varphi: T_1 \to T_2$ be a theory-morphism and A any T_2-equoid. The partial $\Sigma(T_1)$-algebra

$$A\varphi = \big((A_{s\varphi} \mid s \in S(T_1)), ((\sigma\varphi)^A \mid \sigma \in \Omega(T_1))\big)$$

is said to be the *φ-part* of the T_2-equoid A.

If $f: A \to B$ is a homomorphism between T_2-equoids, the $S(T_1)$-indexed family of $S(T_1)$-functions

$$f\varphi = \big(f_{s\varphi}: A_{s\varphi} \to B_{s\varphi} \mid s \in S(T_1)\big)$$

is called the *φ-part* of the T_2-homomorphism $f: A \to B$. ■

Definition 3.5.3 would not fit in our approach if the φ-part of a T_2-equoid were not a T_1-equoid. By the following corollary the preceding definition is justified.

Corollary 3.5.4. *For every theory-morphism $\varphi: T_1 \to T_2$ and for every T_2-equoid A the φ-part is a T_1-equoid. For every T_2-homomorphism $f: A \to B$ the φ-part*

$$f\varphi: A\varphi \to B\varphi$$

is a homomorphism between T_1-equoids.

Proof. First we point out that for every $S(T_1)$-sorted system of variables $v: X \to S(T_1)$ it can be proved easily by induction that

$$t^{(A\varphi)} = (t\varphi)^A$$

holds for every term $t \in T(\Sigma(T_1), v)$.

An application to existence equations and sets of existence equations leads to

$$(A\varphi)_{(v:t_1 \overset{e}{=} t_2)} = A_{(v:t_1 \overset{e}{=} t_2)\varphi}$$

and

$$(A\varphi)_{(v:G)} = A_{(v:G)\varphi}.$$

With these identities we can prove that $A\varphi$ is a $\theta(T_1)$-equoid. Because of

$$\mathfrak{A}(T_2) \vdash \big(w\varphi: (\text{def}\ \sigma)\ \varphi \leftrightarrow \text{def}\ (\sigma\varphi)\big)$$

for every $\sigma \in \Omega(T_1)$, $\sigma: w \to s$, we obtain by Theorem 3.1.9 that

$$A_{(\text{def}\,\sigma)\varphi} = A_{\text{def}(\sigma\varphi)}$$

holds for every T_2-equoid, and so

$$\text{dom}\ \sigma^{(A\varphi)} = \text{dom}(\sigma\varphi)^A = A_{\text{def}(\sigma\varphi)} = A_{(\text{def}\,\sigma)\varphi} = (A\varphi)_{\text{def}\,\sigma}$$

for every $\sigma \in \Omega(T_1)$, i.e. $A\varphi$ is a $\theta(T_1)$-equoid. Now let A by any T_2-equoid and $(v: G \to t_1 \overset{e}{=} t_2) \in \mathfrak{A}(T_1)$ any axiom of T_1. Because of condition (2) of Definition 3.5.2

$$\mathfrak{A}(T_2) \vdash (v\varphi: G\varphi \to t_1\varphi \overset{e}{=} t_2\varphi)$$

holds, which yields by use of Theorem 3.1.9

$$A_{(v:G)\varphi} \subseteq A_{(v:t_1 \overset{e}{=} t_2)\varphi}$$

for every T_2-equoid A. By means of the preceding identities we get

$$(A\varphi)_{(v:G)} \subseteq (A\varphi)_{(v:t_1 \overset{e}{=} t_2)}$$

and so $A\varphi$ satisfies each axiom of T_1, i.e. $A\varphi$ is a T_1-equoid.

Now we assume that $f: A \to B$ is any T_2-homomorphism, $\sigma \in \Omega(T_1)$ with $\sigma: w \to s$, and $\boldsymbol{a} \in \operatorname{dom} \sigma^A$.

Due to

$$\operatorname{dom} \sigma^{(A\varphi)} = \operatorname{dom}(\sigma\varphi)^A$$

we obtain

$$\boldsymbol{a}f_{w\varphi} \in \operatorname{dom}\,(\sigma\varphi)^B = \operatorname{dom} \sigma^{(B\varphi)},$$

so that

$$\boldsymbol{a}(f\varphi)_w = \boldsymbol{a}f_{w\varphi} \in \operatorname{dom} \sigma^{(B\varphi)}.$$

Finally the following holds:

$$\begin{aligned}(\boldsymbol{a}\sigma^{(A\varphi)})\,(f\varphi)_s &= (\boldsymbol{a}\sigma^{(A\varphi)})\,f_{s\varphi} = \big(\boldsymbol{a}(\sigma\varphi)^A\big)\,f_{s\varphi} = (\boldsymbol{a}f_{w\varphi})\,(\sigma\varphi)^B \\ &= (\boldsymbol{a}f_{w\varphi})\,\sigma^{(B\varphi)} = \big(\boldsymbol{a}(f\varphi)_w\big)\,\sigma^{(B\varphi)},\end{aligned}$$

i.e. $f\varphi: A\varphi \to B\varphi$ is a T_1-homomorphism. ■

According to Definition 3.5.2 a hep-theory T_1 is said to be a *subtheory* of a hep-theory T_2 if $S(T_1) \subseteq S(T_2)$, $\Omega(T_1) \subseteq \Omega(T_2)$ and if the inclusions form a theory-morphism, i.e.

$$\mathfrak{A}(T_2) \vdash (w: \operatorname{def}_{T_1}\sigma \leftrightarrow \operatorname{def}_{T_2}\sigma) \quad \text{for every} \quad \sigma \in \Omega(T_1),$$

and

$$\mathfrak{A}(T_2) \vdash (v: G \to t_1 \overset{e}{=} t_2)\,\varphi$$

for every $(v: G \to t_1 \overset{e}{=} t_2) \in \mathfrak{A}(T_1)$.

Also in the case of a subtheory we do not require the inclusion $\mathfrak{A}(T_1) \subseteq \mathfrak{A}(T_2)$ and the identity $\operatorname{def}_{T_1}\sigma = \operatorname{def}_{T_2}\sigma$, although these will be true usually. If T_1 is a subtheory of T_2, this will be denoted briefly by

$$T_1 \subseteq T_2.$$

For the sake of simplicity we do not explicitly use an inclusion morphism in the case of a subtheory, but we use the following notion:

$$A{\downarrow}T_1 \quad \text{and} \quad f{\downarrow}T_1{:}\,A{\downarrow}T_1 \to B{\downarrow}T_1$$

for every T_2-equoid A and every T_2-homomorphism $f: A \to B$, respectively. We speak of the T_1-part $A\downarrow T_1$ of the T_2-equoid A or of the underlying T_1-equoid $A\downarrow T_1$ of the T_2-equoid A.

For instance, the theory of directed graphs

T_{GRAPH} **is sorts** Edges, Nodes
  oprn source(Edges) $\to$ Nodes
    target(Edges) $\to$ Nodes
  axioms none
end T_{GRAPH}

corresponds to a subtheory of the theory $T_{\text{CAT}} = (\Sigma_{\text{CAT}}, \mathfrak{A}^*_{\text{CAT}})$ defined in Section 2.5. This correspondence is given by the theory-morphism

$$\varphi: T_{\text{GRAPH}} \to T_{\text{CAT}}$$

given by

$$\text{Edges}\,\varphi = \text{Mor}, \quad \text{Nodes}\,\varphi = \text{Obj},$$

$$\text{source}\,\varphi = \text{source}, \quad \text{target}\,\varphi = \text{target}.$$

Thus, for every small category $C \in \text{ALG}(T_{\text{CAT}})$ the φ-part $C\varphi$ is a directed graph, where the objects of C are interpreted as the nodes of $C\varphi$ and the morphisms of C are interpreted as the edges of $C\varphi$.

The construction of the φ-part A for a theory-morphism $\varphi: T_1 \to T_2$ and a T_2-equoid A suggests looking for ways of inverting this construction, i.e. looking for ways to construct a T_2-equoid for every given T_1-equoid and every theory-morphism $\varphi: T_1 \to T_2$ in a canonical way.

One way to extend a T_1-equoid B to a T_2-equoid with respect to a theory-morphism $\varphi: T_1 \to T_2$ is given by the following definition.

Definition 3.5.5. Let $\varphi: T_1 \to T_2$ be a theory-morphism. A T_2-equoid F is said to be *freely generated* by a T_1-equoid B *according* to $\varphi: T_1 \to T_2$ if

(1) there is a T_1-homomorphism $h: B \to F\varphi$, such that

(2) for every T_2-equoid A and every T_1-homomorphism $f: B \to A\varphi$ there is exactly one T_2-homomorphism $f^{\#}: F \to A$ with $f = h \circ (f^{\#}\varphi)$. ■

A pictoral representation of the situation is:

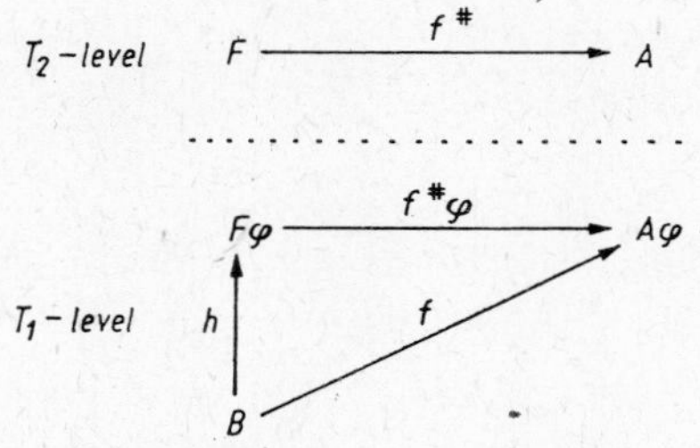

To illustrate this concept we discuss some special cases. If T_1 is the *empty theory* denoted by $T_\emptyset$, then this theory has exactly one model, i.e. there is exactly one $T_\emptyset$-equoid namely the empty sequence of carrier sets together with the empty sequence of fundamental operations. This uniquely determined $T_\emptyset$-equoid will also be denoted by $\emptyset$ so that $\mathrm{ALG}(T_\emptyset) = \{\emptyset\}$. A T-equoid F freely generated by $\emptyset$ according to the inclusion $T_\emptyset \subseteq T$ is then characterized by the fact that for every T-equoid A exactly one T-homomorphism $f: F \to A$ exists. We see that in this special case the notion introduced coincides with the concept of an initial T-equoid.

Now let us assume that $T_0 \subseteq T$ with $S(T_0) = S(T) = S$ and $\Omega(T_0) = \emptyset$. In this case T_0-equoids are exactly S-sets and T_0-homomorphisms are exactly S-functions. A T-equoid freely generated by an S-set $X = (X_s \mid s \in S)$ according to the inclusion $T_0 \subseteq T$ is then characterized by the existence of an S-function $h: X \to F{\downarrow}T_0$ such that for every T-equoid A and every S-function $f: X \to A{\downarrow}T_0$ exactly one T-homomorphism $f^\#: F \to A$ exists with $f = h \circ (f^*{\downarrow}T_0)$. Since the construction $A{\downarrow}T_0$ associates with every T-equoid A its S-carrier and $f^\#{\downarrow}T_0 = f^\#$ the notion introduced coincides with the well-known concept of a T-equoid freely generated by an S-set X.

More generally, the notion introduced allows us to speak of an Abelian group F freely generated by a group G which is implemented by $F = G/K$ where K is the commutator of G.

With respect to partial algebras we can also speak of a small category freely generated by a directed graph. This construction can be implemented by the so-called category of finite paths over a directed graph (see Hasse and Michler (1966)).

From Definition 3.5.5 it follows immediately that a T_2-equoid F freely generated by a T_1-equoid B according to $\varphi: T_1 \to T_2$ is determined uniquely up to isomorphism.

By means of the calculus introduced in Section 3.2 we can prove that for every theory-morphism $\varphi: T_1 \to T_2$ relatively free equoids exist or in terms of the theory of categories that the forgetful functor $A \mapsto A\varphi$, for every T_2-equoid A, has a left-adjoint functor.

Theorem 3.5.6. *For every theory-morphism $\varphi: T_1 \to T_2$ and every T_1-equoid B there exists a T_2-equoid F freely generated by B according to $\varphi: T_1 \to T_2$.*

Proof. In Section 3.3 we proved that any T_1-equoid B is isomorphic to a T_1-equoid $F(\mathfrak{A}(T_1), G, v)$ with an appropriate set of existence equations $(v: G)$, $v: X \to S(T_1)$. Consequently, we can suppose

$$B = F(\mathfrak{A}(T_1), G, v)$$

without loss of generality. Taking

$$F = F(\mathfrak{A}(T_2), G\varphi, v\varphi) \in \mathrm{ALG}(T_2) \quad \text{and} \quad \boldsymbol{e} \in F_{(v:G)}$$

as universal solution we obtain

$$\boldsymbol{e} \in F_{(v:G)\varphi} = (F\varphi)_{(v:G)},$$

due to Proof 3.5.4.

If $\boldsymbol{l} \in F(\mathfrak{A}(T_1), G, v)_{(v:G)}$ denotes the universal solution it follows that there is exactly one homomorphism $h: B \to F\varphi$ with $\boldsymbol{l}h_v = \boldsymbol{e}$. Now let any $A \in \mathrm{ALG}(T_2)$ and $f: B \to A\varphi$ be given. Since a homomorphism preserves solutions of $(v: G)$ we may infer

$$\boldsymbol{l}f_v \in (A\varphi)_{(v:G)} = A_{(v:G)\varphi},$$

so that there is exactly one T_2-homomorphism

$$f^{\#}: F \to A \quad \text{with} \quad \boldsymbol{e}f^{\#}_{v\varphi} = \boldsymbol{l}f_v.$$

This identity leads to

$$\boldsymbol{l}(h \circ (f^{\#}\varphi))_v = \boldsymbol{l}(h_v \circ (f^{\#}\varphi)_v) = \boldsymbol{l}(h_v \circ f^{\#}_{v\varphi}) = (\boldsymbol{l}h_v)\,(f^{\#}_{v\varphi}) = \boldsymbol{e}f^{\#}_{v\varphi} = \boldsymbol{l}f_v$$

and so $h \circ (f^{\#}\varphi) = f$.

If $\hat{f}: F \to A$ is any T_2-homomorphism with $h \circ (\hat{f}\varphi) = f$ we can conclude

$$\boldsymbol{e}f^{\#}_{v} = \boldsymbol{l}f_v = \boldsymbol{l}(h \circ (\hat{f}\varphi))_v = (\boldsymbol{l}h_v)\,(\hat{f}\varphi)_v = \boldsymbol{e}(\hat{f}\varphi)_v = \boldsymbol{e}\hat{f}_{v\varphi}$$

which involves

$$\hat{f} = f^{\#}.$$ ■

We mentioned previously that the construction of the category of paths over a directed graph forms an example for Theorem 3.5.6. To demonstrate this we start with a sorted system of variables

$$w: \{x_1, x_2, x_3, x_4, x_5\} \to \{\text{Edges, Nodes}\}$$

with $x_iw = \text{Edges}$ for $i = 1, 2, 3, 4, 5$ and with a set of existence equations

$$(w: G) = (w: \{\text{source}(x_1) \stackrel{e}{=} \text{source}(x_2), \text{target}(x_2) \stackrel{e}{=} \text{source}(x_3), \text{target}(x_3) \stackrel{e}{=} \text{source}(x_5), \text{source}(x_3) \stackrel{e}{=} \text{source}(x_4), \text{target}(x_5) = \text{target}(x_4)\}).$$

A directed graph G freely generated by $(w: G)$ is graphically described as follows:

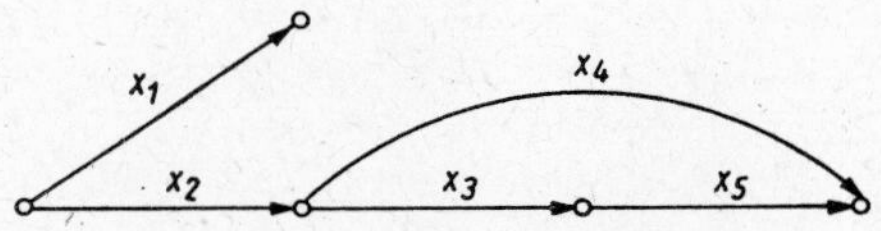

Taking the theory-morphism

$$\varphi: T_{\mathrm{GRAPH}} \to T_{\mathrm{CAT}}$$

as defined above it results in

$$w\varphi\colon \{x_1, \ldots, x_5\} \to \{\text{Obj}, \text{Mor}\}$$

with $x_i w\varphi = \text{Mor}$ for $i = 1, 2, \ldots, 5$ and $(w\colon G)\ \varphi$, so that the small category C freely generated by the directed graph G according $\varphi\colon T_{\text{GRAPH}} \to T_{\text{CAT}}$ is given by

$$C = F(\mathfrak{A}_{\text{CAT}}, G\varphi, w\varphi)$$

which can be represented by the following diagram:

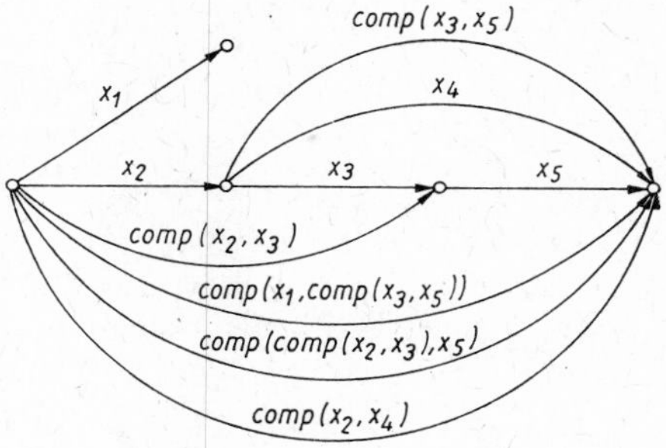

Every additional morphism corresponds to a non-trivial, finite path. In our example there are four such paths.

We conclude this section with a second example. For this reason let the theory T_2 with a subtheory T_1 be given:

T_1 **is sorts** Bool, Elements
  oprn $t, f \to$ Bool
    equal(Elements,Elements) $\to$ Bool
  axioms x,y: Elements
    equal $(x,x) = t$
    if equal$(x,y) = t$ **then** $x = y$
end T_1

T_2 is T_1 **with**
  sorts Sets
  oprn 1 $\to$ Sets
    ap(Sets,Elements) $\to$ Sets
  axioms x,y: Elements, m: Sets
    ap(ap$(m,x),x)$ = ap(m,x)
    ap(ap$(m,x),y)$ = ap(ap$(m,y),x)$
end T_2

This notation of hep-theories, similar to the notation of Chapter 1, differs from the formal definition which requires that every axiom is of the form

$$(v\colon G \to t_1 \overset{e}{=} t_2).$$

First, we can in almost all examples assume that all axioms have the same S-sorted system of variables. With the declaration

$$x, y\colon \text{Elements}, \quad m\colon \text{Sets}$$

we define for instance the following S-sorted system

$$w\colon \{x, y, m\} \to \{\text{Elements,Sets,Bool}\}$$

with $xw = yw = $ Elements and $mw = $ Sets. However, we have not introduced the identifier 'w'.

Secondly, if in $(v\colon G \to t_1 \overset{e}{=} t_2)$ the premise G is empty, we write briefly $t_1 = t_2$ and if G is not empty then we use the notation

$$\textbf{if } G \textbf{ then } t_1 = t_2.$$

Now let us assume that $B \in \text{ALG}(T_1)$ with

$$B_{\text{Bool}} = \{T, F\}, \qquad B_{\text{Elements}} = M$$

any set,

$$t^B = T, \qquad f^B = F,$$

and

$$\text{equal}^B(x, y) = T \quad \text{iff} \quad x = y \quad \text{for all} \quad x, y, \in M.$$

Of course, not every T_1-equoid is of this kind. But, if we choose the T_1-equoid in this kind, then there is a nice description of a T_2-equoid F freely generated by B according to the inclusion $T_1 \subseteq T_2$. We can take

$$F_{\text{Bool}} = B_{\text{Bool}}, \quad F_{\text{Elements}} = B_{\text{Elements}}, \quad t^F = t^B, \quad f^F = f^B,$$

and

$$\text{equal}^F = \text{equal}^B, \quad \text{i.e.} \quad F{\downarrow}T_1 = B.$$

The remaining sort name and the remaining operators of T_2 will be interpreted as follows:

$$F_{\text{Sets}} = \text{set of all finite subsets of } M = F_{\text{Elements}},$$

$$l^F = \emptyset \in F_{\text{Sets}},$$

$$\text{ap}^F(m, x) = m \cup \{x\} \quad \text{for every} \quad m \in F_{\text{Sets}}, \quad x \in F_{\text{Elements}}.$$

By induction on the number of elements of finite subsets one can prove that F is freely generated by B.

The characteristic of this example is that T_2 has a new sort name and no new operators whose range is a sort name of T_1. All new operators generate values of an additional sort. This property of the theory extension $T_1 \subseteq T_2$ guarantees that for every T_2-equoid F freely generated by a T_1-equoid B the identity $B = F \downarrow T_1$ holds. This property of a theory extension will be of great importance for the next chapter, i.e. for the applicability of Theorem 3.5.6 to parameterized abstract data types as described in Chapter 1.

Now we prove a property of relatively free equoids that enables us to build specifications of abstract data types step by step.

Corollary 3.5.7. *Let* $\varphi: T_1 \to T_2$, $\psi: T_2 \to T_3$ *be theory-morphisms. If* $F \in \mathrm{ALG}(T_2)$ *is freely generated by* $B \in \mathrm{ALG}(T_1)$ *according to* $\varphi: T_1 \to T_2$, *and if* $F^\# \in \mathrm{ALG}(T_3)$ *is freely generated by* F *according to* $\psi: T_2 \to T_3$, *then* $F^\#$ *is freely generated by* B *according to* $\varphi \circ \psi: T_1 \to T_3$.

Proof. Let $h: B \to F\varphi$, $g: F \to F^\#\psi$ be the homomorphisms corresponding to Definition 3.5.5. Then

$$h \circ (g\varphi): B \to (F^\#\psi)\,\varphi = F^\#(\varphi \circ \psi).$$

If $A \in \mathrm{ALG}(T_3)$ and $f: B \to A(\varphi \circ \psi) = (A\psi)\,\varphi$, then there is exactly one homomorphism

$$f^\#: F \to A\psi \quad \text{with} \quad h \circ (f^\#\varphi) = f.$$

This involves the existence of exactly one homomorphism

$$f^{\#\#}: F^\# \to A \quad \text{with} \quad g \circ (f^{\#\#}\psi) = f^\#$$

so that

$$f^\#\varphi = \big(g \circ (f^{\#\#}\psi)\big)\,\varphi = (g\varphi) \circ \;(f^{\#\#}\psi)\,\varphi$$

and therefore

$$\big(h \circ (g\varphi)\big)\,\big(f^{\#\#}(\varphi \circ \psi)\big) = h \circ (g\varphi)\,\big((f^{\#\#}\psi)\,\varphi\big) = h \circ (f^\#\varphi) = f. \qquad \blacksquare$$

We illustrate the application of Corollary 3.5.7 by the following extension $T_2 \subseteq T_3$:

> T_3 **is** T_2 **with**
>   **oprn** el(Elements,Sets) → Bool
>   **axioms** x,y: Elements, m: Sets
>   el(x,1) = f
>   **if** el(x,m) = f **then** el(y,ap(m,x)) = equal(x,y)
>   **if** el(x,m) = t **then** el(x,ap(m,y)) = t
> **end** T_3

Let $B \in \mathrm{ALG}(T_1)$, $F \in \mathrm{ALG}(T_2)$ be as above. A T_3-equoid $F^\#$ freely generated by F according $T_2 \subseteq T_3$ is characterized by

$$F^\# \downarrow T_2 = F \quad \text{and} \quad \mathrm{el}^{F^\#} = \text{element relation}.$$

By Corollary 3.5.7 $F^\#$ is also freely generated by B according $T_1 \subseteq T_3$.

4. Canons — initially restricting algebraic theories

The algebraic approach developed in Chapter 2 and 3 is not sufficient for descriptions of those classes of partial algebras that appear as model classes of specifications in Chapter 1. In particular the initial satisfaction of axioms as used for enrichments by definition cannot be expressed by the algebraic notions developed in the preceding two chapters. It is the aim of this chapter to introduce suitable notions, so-called initial restrictions in algebraic theories.

In the first section we define the concept of a *canon* and investigate classes of models of canons. In a slightly restricted form the notion of a canon was introduced in Kaphengst and Reichel (1971). From an intermediate form in Reichel (1979) the final form was defined in Reichel (1980) following a suggestion of Kaphengst. Just the same concept was used in Burstall and Goguen (1980) for the definition of the semantics of the specification language CLEAR. Later Ehrig, see Ehrig (1981), used an extension of this concept for dealing with parameterized data types. But it is worth mentioning here that the full expressive power of the canon concept can only be achieved by the use of partial algebras. This will be made comprehensible in Section 4.2 which is aimed at the development of a formalization of the concept of an abstract algorithm over a parameterized abstract data type. Beside a justification of this notion we will prove a necessary condition for a function over a parameterized abstract data type to be an algorithm, i.e. a computable parameterized function. Additionally we investigate uniquely specifiable functions over parameterized abstract data types. The difference between these two concepts is demonstrated in Chapter 1 by the binary relation in the set of nodes of a directed graph indicating whether two nodes are connected by a finite directed path. This relation is an example of a uniquely specifiable but not computable function.

In the last section of this chapter we introduce morphisms between canons and prove some of their properties. This finishes the set of formal notions necessary for the semantics of the specifications given in Chapter 1. Since the introduction of a specification language is not the aim of this book we will not give a complete and formal definition of the semantics. But there are no serious problems for this task since we have made available all of the tools for inductive definitions even if the production rules, i.e. the generating constructions, are restricted by context conditions.

4.1. Canon model classes

The essential difference between model classes of $(\theta, \mathfrak{A})$-algebras, where θ is an ep-signature and $\mathfrak{A}$ a stable set of elementary implications, and model classes of specifications, containing enrichments by definitions, is demonstrated by the simple specification NAT of Section 1.1. Whereas classes of $(\theta, \mathfrak{A})$-algebras can naturally be axiomatized by first-order theories, it is well known that the model class of NAT cannot be described by a first-order theory. This is because restrictions of model classes by initial satisfaction of elementary implications cannot be expressed in first-order theories. Thus, the algebraic basic notion for specifications, exceeding first-order theories, is given by the following definition of an initial restriction in an algebraic theory, for which Theorem 3.5.6 is the key.

Definition 4.1.1. Let algebraic theories T, T_2, a subtheory T_1 of T_2, and a theory-morphism $\varphi: T_2 \to T$ be given (see Definition 3.5.2). The ordered pair

$$(T_1, \varphi: T_2 \to T)$$

is called an *initial restriction* in T. A T-algebra A satisfies this initial restriction, denoted by

$$A \models (T_1, \varphi: T_2 \to T)$$

if the φ-part $A\varphi$ is freely generated by $(A\varphi)\downarrow T_1$ according to the inclusion $T_1 \subseteq T_2$ (see Definition 3.5.5). ■

We draw attention to the following fact. If

$$A \models (T_1, \varphi: T_2 \to T)$$

then $A\varphi$ is not only freely generated by $(A\varphi)\downarrow T_1$ in the sense of Definition 3.5.5 but $A\varphi$ is a free T_2-*extension* of $(A\varphi)\downarrow T_1$ with respect to the inclusion $T_1 \subseteq T_2$. With respect to an inclusion $T_1 \subseteq T_2$ we say that a T_2-algebra F is a free T_2-extension of a T_1-algebra B if

(1) $F\downarrow T_1 = B$ and
(2) for every $A \in \mathrm{ALG}(T_2)$ and every homomorphism $f: B \to A\downarrow T_1$ there exists exactly one homomorphism $f^{\#}: F \to A$ with $f = f^{\#}\downarrow T_1$.

Let us consider two extremal cases. The first is given by the initial restriction $(T, \mathrm{id}: T \to T)$. This initial restriction in T is trivial since every T-algebra satisfies it. On the other hand, the initial restriction

$$(\emptyset, \mathrm{id}\, T: T \to T)$$

where $\emptyset$ denotes the empty subtheory, is called the *total* initial restriction of T. It is easy to see that exactly the initial T-algebras satisfy the total initial restriction.

An initial restriction $(T_1, \varphi: T_2 \to T)$ is said to be *inner* if T_2 is a subtheory of T and $\varphi: T_2 \to T$ is the inclusion. Almost all realistic examples of initial restrictions, especially all those used in this book, are inner ones. But from a more theoretical point of view it is not possible to restrict the approach to inner initial restrictions.

Since a specification may contain several enrichments by definition we need the following notion.

Definition 4.1.2. An ordered pair

$$(T, \Delta)$$

consisting of a finite algebraic theory T and a finite set Δ of initial restrictions in T is called a *canon*. A T-algebra A is said to be a (T, Δ)*-algebra* if A satisfies every initial restriction of Δ. $\mathrm{ALG}(T, \Delta)$ denotes the class of all (T, Δ)-algebras. ■

Immediately we obtain

$$\mathrm{ALG}(T, \emptyset) = \mathrm{ALG}(T)$$

and $\mathrm{ALG}(T, \{(\emptyset, \mathrm{id}\ T: T \to T)\})$ = isomorphism class of the initial T-algebra. A canon (T, Δ) is called *totally restricted* if

$$\mathrm{ALG}(T, \Delta) = \mathrm{ALG}(T, \{(\emptyset, \mathrm{id}\ T\colon T \to T)\}).$$

In view of this fact we can use totally restricted canons as mathematical representations of abstract data types.

As an illustration of the notion introduced we give the canon described by the specification V-D-GRAPH. For that reason we consider the following theories:

$T_{\text{D-GRAPH}}$ **is sorts** Edges, Nodes
  oprn begin(Edges) → Nodes
    end(Edges) → Nodes
  axioms none

T_{NAT} **is sorts** Nat, Bool
  oprn true, false → Bool
    zero → Nat
    succ(Nat) → Nat
    eq(Nat,Nat) → Bool
  axioms x,y: Nat
    eq(x,x) = true
    eq(x,y) = eq(y,x)
    eq(zero,succ(x)) = false
    eq(succ(x),succ(y)) = eq(x,y)

T_2 **is** $T_{\text{D-GRAPH}}$ **with** T_{NAT}

T_3 **is** T_2 **with**
  sorts E-Sets, N-Sets
  oprn e-empty → E-Sets
    n-empty → N-Sets
    e-join(E-Sets,Edges) → E-Sets
    n-join(N-Sets,Nodes) → N-Sets

axioms x_1,x_2: Edges, y_1,y_2: Nodes, m: E-Sets, k: N-Sets
e-join(e-join(m,x_1),x_1) = e-join(m,x_1)
n-join(n-join(k,y_1),y_1) = n-join(k,y_1)
e-join(e-join(m,x_1),x_2) = e-join(e-join(m,x_2),x_1)
n-join(n-join(k,y_1),y_2) = n-join(n-join(k,y_2),y_1)

$T_{\text{V-D-GRAPH}}$ **is** T_3 **with**
sorts no new sorts
oprn edges $\to$ E-Sets
nodes $\to$ N-Sets
valuation(Edges) $\to$ Nat
axioms x: Edges, y: Nodes
e-join(edges,x) = edges
n-join(nodes,y) = nodes

The specification V-D-GRAPH describes a canon $\mathfrak{C}_{\text{V-D-GRAPH}}$ with two initial restrictions in $T_{\text{V-D-GRAPH}}$:

$$\mathfrak{C}_{\text{V-D-GRAPH}} = (T_{\text{V-D-GRAPH}}, \{(\emptyset, i_1: T_{\text{NAT}} \to T_{\text{V-D-GRAPH}}), (T_2, i_2: T_3 \to T_{\text{V-D-GRAPH}})\})$$

where i_1, i_2 denote the inclusions.

In the following $\mathfrak{C}_{\times\times\times} = (T_{\times\times\times}, \Delta_{\times\times\times})$ will denote the canon that corresponds to the specification $\times\times\times$, $T_{\times\times\times}$ denotes the algebraic theory of $\mathfrak{C}_{\times\times\times}$, and $\Delta_{\times\times\times}$ denotes the set of initial restrictions in $\mathfrak{C}_{\times\times\times}$, corresponding to the enrichments by definition of the specification $\times\times\times$.

The first initial restriction defines the abstract data type of natural numbers and the abstract data type of truth values together with the equality of natural numbers. The second initial restriction defines the parameterized abstract data types of the sets of finite subsets of the sets of edges and nodes, respectively. The second initial restriction represents a parameterized abstract data type because the subtheory T_2 is not totally restricted by the first initial restriction.

The study of properties of model classes of canons is not only a natural algebraic question but also of some interest from the point of view of applications. Based on those properties one can derive tests for deciding if certain constructions of sets of functions can be specified by the initial canon specification method. By virtue of the following result we can state for instance that the general powerset construction cannot be specified and that the classe of all complete lattices is not the model class of any canon.

In the following we present some results obtained by K. Benecke in his thesis for special kinds of canons and extended to general canons in Benecke (1983). We conclude this section with a new result concerning the closure under quotient constructions.

Because a model class of a canon can be the isomorphism class of an initial T-algebra model classes of canons are evidently not necessarily closed with respect to direct products.

By the following example we demonstrate that a class of models of a canon is in general not closed with respect to the intersection of subalgebras.

EXAMPLE 4.0 **is** BOOL **with requirement**
  sorts M
  oprn eq(M,M) $\rightarrow$ Bool
  axioms x,y: M
    eq(x,x) = true
    if eq(x,y) = true **then** $x = y$
end EXAMPLE 4.0

EXAMPLE 4.1 **is** EXAMPLE 4.0 **with definition**
  sorts Q
  oprn $q(M,M) \rightarrow Q$
  axioms x,y,z: M
    if eq(x,y)=false, eq(x,z)=false, eq(y,z)=false
      then $q(x,y){=}q(x,x)$, $q(x,x){=}q(z,z)$ **fi**
end EXAMPLE 4.1

In this specification we use a non-elementary implication as equivalent to two elementary implications with identical premises. This specification describes a canon with two inner initial restrictions. The first initial restriction is given by the subspecification BOOL and the second initial restriction is given by the last enrichment by definition. This enrichment by definition implies that for every algebra

$$A \in \mathrm{ALG}(\mathbb{C}_{\mathrm{EXAMPLE\,4.1}})$$

A_Q is a duplicate of $A_M \times A_M$ provided A_M has no more than two elements. If A_M has three or more elements the implication involves that A_Q consists of exactly one element. Consequently, each algebra A of $\mathrm{ALG}(\mathbb{C}_{\mathrm{EXAMPLE4.1}})$ is up to isomorphism uniquely determined by the set A_M.

Let algebras A, B, $C \in \mathrm{ALG}(\mathbb{C}_{\mathrm{EXAMPLE\,4.1}})$ be given with

$$A_M = \{1, 2, 3, 4\}, \qquad A_Q = \{1\},$$

$$B_M = \{1, 2, 3\}, \qquad C_M = \{1, 2, 4\}, \qquad B_Q = C_Q = A_Q.$$

Then B and C are both subalgebras of A but the subalgebra

$$D = B \cap C$$

of A does not satisfy the second initial restriction, because $D_M = \{1, 2\}$ and $D_Q = \{1\}$. D would only be a model of the canon $\mathbb{C}_{\mathrm{EXAMPLE\,4.1}}$ if it is isomorphic to

$$D_Q = \{(1,1), (2,2), (1,2), (2,1)\}.$$

We see that model classes of canons have a different behaviour from varieties and quasi-varieties. Roughly speaking one can say they have opposite behaviour. The preceding example V-D-GRAPH demonstrates that they behave also in a different way from model classes of first-order theories. Contrary to the cardinality theorem any model of V-D-GRAPH is countable and no model of greater cardinality exists.

Because of the following theorem every model class of a canon has a model of at most countable cardinality if at least one model exists. So we can say that model classes of canons satisfy a downward cardinality theorem.

Theorem 4.1.3. *Let the following be given: any canon (T, Δ), any (T, Δ)-algebra A, and any S-subset $M = (M_s \mid s \in S)$ of the S-carrier of A, where S denotes the set of sort names of T. Then there exists a (T, Δ)-algebra B which is a subalgebra of A so that the cardinality of B_s is less than or equal to the maximum of the cardinality of M_s and of the set of natural numbers for every $s \in S$.*

Proof. If $\Delta = \emptyset$, i.e. if there are no initial restrictions, then we can construct B as the T-subequoid of A generated by M.

For the sake of simplicity we suppose that Δ consists of only one initial restriction $(T_1, \varphi\colon T_2 \to T)$. In the general case one can use the same idea of the following proof, but this idea would then be hidden by notational problems.

Now let be $A \in \mathrm{ALG}(T, \Delta)$ and $M \subseteq A$. Then we abbreviate

$$A'' = A\varphi \in \mathrm{ALG}(T_2),$$
$$A' = A''{\downarrow}T_1 = (A\varphi){\downarrow}T_1 \in \mathrm{ALG}(T_1)$$

so that A'' is freely generated by A'. Without loss of generality we may suppose that the carriers of A are pairwise disjoint. By

$$\hat{w}\colon \cup \{A'_s \mid s \in S(T_1)\} \to S(T_1)$$

and

$$w\colon \cup \{A'_s \mid s \in S(T_2)\} \to S(T_2) \supseteq S(T_1)$$

we denote the sorted systems of variables corresponding to A' taken for an $S(T_1)$-set and $S(T_2)$-set, respectively. Further let $\boldsymbol{e} \in A''_w$ and $\boldsymbol{e} \in A'_{\hat{w}}$ denote the identical inclusion of the $S(T_2)$-set A' into A'' and the identity of the $S(T_1)$-set A', respectively. Corresponding to Section 3.3 we set

$$(\hat{w}\colon G_{A'}) = \{(\hat{w}\colon a \overset{\mathrm{e}}{=} t) \mid t \in T(\Sigma(T_1), \hat{w})_s,\ s \in S(T_1),\ a = \boldsymbol{e}t^{A'}\},$$

$$(w\colon G_{A'}) = \{(w\colon a \overset{\mathrm{e}}{=} t) \mid t \in T(\Sigma(T_1), w_s),\ s \in S(T_1),\ a = \boldsymbol{e}t^{A'}\}.$$

Because of $A \in \mathrm{ALG}(T, \Delta)$ we obtain by Theorem 3.5.6 that

$$A' \text{ is isomorphic to } F\big(\mathfrak{A}(T_1), G_{A'}, \hat{w}\big)$$

and

$$A'' \text{ is isomorphic to } F\big(\mathfrak{A}(T_2), G_{A'}, w\big)$$

and that $\boldsymbol{e} \in A''_w$ is the universal solution of $(w\colon G_{A'})$ in A''.
For every $t \in T\big(\Sigma(T_2), w\big)$ there exists a smallest $S(T_2)$-sorted subsystem

$$w_t\colon \mathrm{var}(t) \to S(T_2)$$

of w with $t \in T\big(\Sigma(T_2), w_t\big)$, where $\mathrm{var}(t)$ is the finite $S(T_2)$-set of all variables used syntactically in the term t. Analogously, for every finite set $(w\colon G)$ of existence equations on w there exists a smallest finite $S(T_2)$-sorted subsystem

$$w_G\colon \mathrm{var}(w\colon G) \to S(T_2)$$

of w with

$$t_1, t_2 \in T\big(\Sigma(T_2), w_G\big)$$

for every $(w\colon t_1 \overset{\mathrm{e}}{=} t_2) \in (w\colon G)$.

According to Definition 3.1.7, for every $(w\colon t_1 \overset{\mathrm{e}}{=} t_2)$ with

$$\mathfrak{A}(T_2) \models (w\colon G_{A'} \to t_1 \overset{\mathrm{e}}{=} t_2)$$

there exists a finite subset of $G_{A'}$, denoted by $G_{(w:t_1 \overset{\mathrm{e}}{=} t_2)}$, with

$$\mathfrak{A}(T_2) \models (w\colon G_{(w:t_1 \overset{\mathrm{e}}{=} t_2)} \to t_1 \overset{\mathrm{e}}{=} t_2).$$

After these notational conventions we can start with the construction of the required (T, Δ)-algebra B, which all be constructed by iteration of three different closure operations starting with M.

For each natural number $k \in \mathbb{N}$ we set

(o) $M^0 = M$,

(i) $M^{3k+1} = M^{3k} \cup \big\{\mathrm{var}(t) \mid t \in T\big(\Sigma(T_2), w\big)_s, s \in S(T_2), s \notin S(T_1), \boldsymbol{e}t^{A''} \in M_s^{3k}\big\}$,

(ii) $M^{3k+2} = M^{3k+1} \cup \big\{\mathrm{var}(G_{(w:t_1 \overset{\mathrm{e}}{=} t_2)}) \mid t_1, t_2 \in T\big(\Sigma(T_2), w\big),$
$\boldsymbol{e} \in A_{(w:t_1 \overset{\mathrm{e}}{=} t_2)}, \mathrm{var}(t_1) \subseteq M^{3k+1}, \mathrm{var}(t_2) \subseteq M^{3k+1}\big\}$,

(iii) $M^{3k+3} = M^{3k+2} \cup \{\boldsymbol{m}\sigma^A \mid \sigma \in \Omega(T), \sigma\colon v \to s, \boldsymbol{m} \in M_v^{3k+2}\}$

and finally

$$B = \bigcup_{k \in \mathbb{N}} M^{3k+3}.$$

Note that these constructions are performed with $S(T)$-sets. For the general proof, where Δ consists of more than one initial restriction in T, we would have to modify the closure operations (i) and (ii). For every initial restriction of Δ there would be an expression analogous to the one which we have constructed.

Due to the closure operation (iii) the $S(T)$-subset B is closed with respect to all fundamental operations σ^A, $\sigma \in \Omega(T)$, and so B is the carrier of a T-subequoid of A.

It remains to show that B satisfies the initial restriction

$$(T_1, \varphi: T_2 \to T).$$

Just as above we abbreviate

$$B'' = B\varphi \in \mathrm{ALG}(T_2),$$
$$B' = (B\varphi)\downarrow T_1 \in \mathrm{ALG}(T_1)$$

and denote by

$$\hat{u}: \cup \{B'_s \mid s \in S(T_1)\} \to S(T_1),$$
$$u: \cup \{B'_s \mid s \in S(T_1)\} \to S(T_2) \supseteq S(T_1)$$

the sorted systems of variables corresponding to B taken for an $S(T_1)$-set and $S(T_2)$-set, respectively. The construction of B implies that $\hat{u}$ is a subsystem of $\hat{w}$ and u is a subsystem of w. Let $\boldsymbol{i} \in B''_u$ and $\boldsymbol{i} \in B'_{\hat{u}}$ denote the identical inclusion of the $S(T_1)$-set B' into B'' and the identity of the $S(T_1)$-set B', respectively. Finally we set

$$(\hat{u}: G_{B'}) = \{(\hat{u}: b \overset{e}{=} t) \mid t \in T(\Sigma(T_1), \hat{u})_s, s \in S(T_1), b = \boldsymbol{i}t^{B'}\},$$
$$(u: G_{B'}) = \{(u: b \overset{e}{=} t) \mid t \in T(\Delta(T_1), u)_s, s \in S(T_1), b = \boldsymbol{i}t^{B'}\}.$$

Contrary to the first part of this proof we have now to show that B'' is isomorphic to $F(\mathfrak{A}(T_2), G_{B'}, u)$.

According to Theorem 3.2.4 for the solution

$$\boldsymbol{i} \in B''_{(u:G_{B'})}$$

there exists exactly one homomorphism

$$i^{\#}: F(\mathfrak{A}(T_2), G_{B'}, u) \to B''$$

defined by

$$[t, G_{B'}]\, \boldsymbol{i}^{\#} = \boldsymbol{i}t^{B''}$$

for every $[t, G_{B'}] \in F(\mathfrak{A}(T_2), G_{B'}, u)$.

We will prove first that the homomorphism $i^{\#}$ is surjective and second that it is injective, too. By Theorem 2.4.2 then follows that $i^{\#}$ is an isomorphism.

If $b \in B''_s$ and $s \in S(T_1)$ then b itself is a variable of the $S(T_2)$-sorted system of variables u and so

$$[b, G_{B'}]\, i^{\#} = b,$$

i.e. $i^{\#}_s$ is surjective.

If $b \in B''_s \subseteq A''_s$ and $s \in S(T_2)$, $s \notin S(T_1)$ then according to the closure operation (i) there exists a term $t \in T(\Sigma(T_2), w)_s$ with $b = \boldsymbol{e}t^{A''}$ and $\mathrm{var}(t) \subseteq B$

such that

$$b = et^{A''} = it^{B''}.$$

For the proof of the injectivity let us assume

$$[t_1, G_{B'}]\, i^{\#} = [t_2, G_{B'}]\, i^{\#}.$$

This implies

$$it_1^{B''} = it_2^{B''} \quad \text{and} \quad et_1^{A''} = et_2^{A''}.$$

Owing to the closure operation (ii) a finite subset $G_{(w:t_1 \overset{e}{=} t_2)}$ of G_A, exists with

$$\mathfrak{A}(T_2) \models (w\colon G_{(w:t_1 \overset{e}{=} t_2)} \rightarrow t_1 \overset{e}{=} t_2)$$

and $\mathrm{var}(G_{(w:t_1 \overset{e}{=} t_2)}) \subseteq B$.

This last relation involves that $G_{(w:t_1 \overset{e}{=} t_2)}$ may be regarded as a set of existence equations on the $S(T_2)$-sorted system of variables u, and so

$$(w\!\downarrow\! u\colon G_{(w:t_1 \overset{e}{=} t_2)})$$

is a finite subset of G_B, and

$$\mathfrak{A}(T_2) \models (w\!\downarrow\! u\colon G_{(w:t_1 \overset{e}{=} t_2)} \rightarrow t_1 \overset{e}{=} t_2).$$

This implies

$$[t_1, G_{B'}] = [t_2, G_{B'}]$$

so that the homomorphism

$$i^{\#}\colon F(\mathfrak{A}(T_2), G_{B'}, u) \rightarrow B''$$

is injective.

The cardinality of B_s for every $s \in S(T)$ cannot be greater than the maximum of the cardinality of M_s and $\mathbb{N}$, because $\mathrm{var}(t)$ is a finite $S(T)$-sorted set for every term t, $\Omega(\mathrm{G})$ is a finite set of operators, and Δ is a finite set of initial restrictions. ∎

By means of Theorem 4.1.3 we can prove that some classes of algebraic structures cannot be described as a class of models of any canon.

Let us consider for instance the following theory:

SETS **is sorts** Elements, Sets
  oprn ∅ → Sets
    add(Sets,Elements) → Sets
  axioms x,y: Elements, s: Sets
    add(add(s,x),x) = add(s,x)
    add(add(s,x),y) = add(add(s,y),x)

and the class

$$\mathfrak{S} = \{A \in \mathrm{ALG(SETS)} \mid A_{\mathrm{Sets}} = \text{set of all subsets of } A_{\mathrm{Elements}},\ \emptyset^A = \text{empty subset of } A_{\mathrm{Elements}}, \text{ and } \mathrm{add}^A(s, x) = s \cup \{x\}\}.$$

Then there is no canon (T, Δ) with $\mathrm{ALG}(T, \Delta) = \mathfrak{S}$, since for the algebra $A \in \mathfrak{S}$ with $A_{\mathrm{Elements}} = \mathbb{N}$ and the two-sorted subset M with $M_{\mathrm{Elements}} = \mathbb{N}$ and $M_{\mathrm{Sets}} = \emptyset$ there is no subalgebra B of A with $B \in \mathfrak{S}$ and $M \subseteq B$ so that B_{Sets} is of countable cardinality. $B \in \mathfrak{S}$ implies that B_{Sets} is of uncountable cardinality. This example proves that the usual power-set construction cannot be specified by the initial canon specification method.

In the same way one can see that the class of all continuous fields can also not be specified in terms of initial canons. The continuous field of real numbers does not contain a countable continuous subfield. However, the class of all fields can easily be specified by means of a suitable canon.

A further example of a class not specifiable as model class of a canon is the class of all complete lattices. To see this let us consider the complete lattice A of all sets of natural numbers and let us take $M \subseteq A$ as the countable set of all atoms and anti-atoms of A. Once more one can prove that no complete countable sublattice of A exists that includes M.

The following property of model classes of canons is most suitably representable in terms of the theory of categories (see MACLANE (1971)), namely as closure property with respect to special colimits. Because we do not suppose that readers are familiar with these concepts we will introduce them here.

Let G be any finite directed graph and let (T, Δ) be any canon. A *diagram*

$$\Phi: G \to \mathrm{ALG}(T, \Delta)$$

in the class of models of the canon (T, Δ) is given by two functions

$$\Phi_{\mathrm{Nodes}}: G_{\mathrm{Nodes}} \to \mathrm{ALG}(T, \Delta)$$

$$\Phi_{\mathrm{Edges}}: G_{\mathrm{Edges}} \to \text{class of all homomorphism between } (T, \Delta)\text{-algebras}$$

so that for every $k \in G_{\mathrm{Edges}}$ with $n_1 = k\,\mathrm{begin}^G$ and $n_2 = k\,\mathrm{end}^G$

$$k\Phi_{\mathrm{Edges}}: n_1\Phi_{\mathrm{Nodes}} \to n_2\Phi_{\mathrm{Nodes}}$$

is a homomorphism from the image of n_1 to the image of n_2.

Let $\Phi: G \to \mathrm{ALG}(T, \Delta)$ be any diagram in $\mathrm{ALG}(T, \Delta)$. A T-equoid L together with a family

$$\big(f(n): n\Phi_{\mathrm{Nodes}} \to L \mid n \in G_{\mathrm{Nodes}}\big)$$

of homomorphisms is called a *colimit* of Φ in $\mathrm{ALG}(T, \Delta)$ if

(1) $(k\Phi_{\mathrm{Edges}}) \circ f(n_2) = f(n_1)$ for every edge $k \in G_{\mathrm{Edges}}$ directed from the node n_1 to the node n_2;

(2) For every family

$$\big(g(n)\colon n\Phi_{\text{Nodes}} \to A \mid n \in G_{\text{Nodes}}\big)$$

of homomorphisms with $A \in \text{ALG}(T)$ and $(k\Phi_{\text{Edges}}) \circ g(n_2) = g(n_1)$ for every edge $k \in G_{\text{Edges}}$ directed from n_1 to n_2 there exists exactly one homomorphism

$$h\colon L \to A$$

with $f(n) \circ h = g(n)$ for every $n \in G_{\text{Nodes}}$.

All so-called *quotient constructions* dealt with in Section 3.3 are constructions of colimits in $\text{ALG}(T)$ for special kinds of diagrams. Using the terminology just introduced the results of Section 3.3 can be summarized by the statement that for every finite diagram a colimit exists in $\text{ALG}(T)$. It is evident that a colimit is determined uniquely up to isomorphism by the diagram

We say that a canon model class $\text{ALG}(T, \Delta) \subseteq \text{ALG}(T)$ is *closed with respect to a colimit* of a diagram $\Phi\colon G \to \text{ALG}(T, \Delta)$ if the existing colimit L of $\Phi\colon G \to \text{ALG}(T, \Delta)$ in $\text{ALG}(T)$ is even in $\text{ALG}(T, \Delta)$.

We say that an initial restriction

$$r = (T_1, \varphi\colon T_2 \to T)$$

in T *preserves a colimit* $L \in \text{ALG}(T)$ of $\Phi\colon G \to \text{ALG}(T, \Delta)$ if at first $L\varphi$ is a colimit of the diagram

$$\Phi\varphi\colon G \to \text{ALG}(T_2)$$

defined by

$$n(\Phi\varphi) = (n\Phi_{\text{Nodes}})\,\varphi \quad \text{for every node} \quad n \in G_{\text{Nodes}},$$

$$k(\Phi\varphi) = (k\Phi_{\text{Edges}})\,\varphi \quad \text{for every edge} \quad K \in G_{\text{Edges}},$$

and if secondly

$$(L\varphi){\downarrow}T_1$$

is a colimit in $\text{ALG}(T_1)$ of the diagram

$$\Phi\varphi{\downarrow}T_1\colon G \to \text{ALG}(T_1)$$

defined by

$$n(\Phi\varphi{\downarrow}T_1) = \big((n\Phi_{\text{Nodes}})\,\varphi\big){\downarrow}T_1 \quad \text{for every} \quad n \in G_{\text{Nodes}},$$

$$k(\Phi\varphi{\downarrow}T_1) = \big((k\Phi_{\text{Edges}})\,\varphi\big){\downarrow}T_1 \quad \text{for every} \quad k \in G_{\text{Edges}}.$$

Now we are able to state a further closure property of canon model classes.

Theorem 4.1.4. *A canon model class* $\text{ALG}(T, \Delta)$ *is closed with respect to those colimits that are preserved by every initial restriction of* Δ.

Proof. Let the following be given: a diagram $\Phi\colon G \to \text{ALG}(T, \Delta)$, a colimit

$$\big(f(n)\colon n\Phi_{\text{Nodes}} \to L \mid n \in G_{\text{Nodes}}\big)$$

of Φ in ALG(T), and any initial restriction (T_1, φ: $T_2 \to T$) of Δ. Then we introduce the following abbreviations:

$$f''(n) = f(n)\,\varphi,$$
$$f'(n) = \big(f(n)\big)\,\varphi\!\downarrow\! T_1$$

for every $n \in G_{\text{Nodes}}$.

Now we have to prove that $L\varphi$ is freely generated by $(L\varphi)\!\downarrow\! T_1$, i.e. we have to show that for each $B \in \text{ALG}(T_2)$ and every homomorphism

$$h\colon (L\varphi)\!\downarrow\! T_1 \to B\!\downarrow\! T_1$$

exactly one homomorphism

$$h^{\#}\colon L\varphi \to B$$

with $h = h^{\#}\!\downarrow\! T_1$ exists. Because $n\Phi_{\text{Nodes}} \in \text{ALG}(T, \Delta)$ for every $n \in G_{\text{Nodes}}$, the equoid $n(\Phi\varphi) \in \text{ALG}(T_2)$ is freely generated by $n(\Phi\varphi\!\downarrow\! T_1) \in \text{ALG}(T_1)$, so that for every $n \in G_{\text{Nodes}}$ exactly one homomorphism

$$h'(n)\colon n(\Phi\varphi) \to B$$

exists with $h'(n)\!\downarrow\! T_1 = f''(n) \circ h$.

Since (T_1, φ: $T_2 \to T$) preserves the given colimit, there is exactly one homomorphism

$$h^{\#}\colon L\varphi \to B$$

with $f'(n) \circ h^{\#} = h'(n)$ for every $n \in G_{\text{Nodes}}$.

It remains to prove that $h^{\#}\!\downarrow\! T_1 = h$ and that $h^{\#}$ is uniquely determined by this condition. Using the equalities $f'(n) \circ h^{\#} = h'(n)$, $h'(n)\!\downarrow\! T_1 = f''(n) \circ h$, and $f''(n) = f'(n)\!\downarrow\! T_1$ for every $n \in G_{\text{Nodes}}$ we obtain

$$\begin{aligned} f''(n) \circ (h^{\#}\!\downarrow\! T_1) &= \big(f'(n)\!\downarrow\! T_1\big) \circ (h^{\#}\!\downarrow\! T_1) = \big(f'(n) \circ h^{\#}\big)\!\downarrow\! T_1 = h'(n)\!\downarrow\! T_1 \\ &= f''(n) \circ h \end{aligned}$$

for every $n \in G_{\text{Nodes}}$ which implies $h = h^{\#}\!\downarrow\! T_1$ since

$$\big(f''(n)\colon n(\Phi\varphi\!\downarrow\! T_1) \to (L\varphi)\!\downarrow\! T_1 \mid n \in G_{\text{Nodes}}\big)$$

is a colimit of

$$\Phi\varphi\!\downarrow\! T_1\colon G \to \text{ALG}(T_1).$$

If $h^{\#\#}\colon L\varphi \to B$ is any homomorphism with $h^{\#\#}\!\downarrow\! T_1 = h$ we obtain

$$\big(f'(n) \circ h^{\#\#}\big)\!\downarrow\! T_1 = f''(n) \circ h$$

for every $n \in G_{\text{Nodes}}$ and so

$$h'(n) = f'(n) \circ h^{\#} = f'(n) \circ h^{\#\#}$$

for every $n \in G_{\text{Nodes}}$ because $n(\Phi\varphi)$ is freely generated by $n(\Phi\varphi\!\downarrow\! T_1)$ for every $n \in G_{\text{Nodes}}$.

This leads to $h^{\#\#} = h^{\#}$ since

$$\big(f'(n)\colon n(\Phi\varphi) \to L\varphi \mid n \in G_{\text{Nodes}}\big)$$

is a colimit of $\Phi\varphi\colon G \to \mathrm{ALG}(T_2)$. ■

If we apply the introduced terminology of colimits to Corollary 3.4.4 we can say that for every theory T and every diagram

$$\Phi\colon G \to \mathrm{ALG}(T)$$

where G is a not necessarily finite chain, the $S(T)$-carrier of a colimit of Φ in $\mathrm{ALG}(T)$ is a colimit of the diagram

$$\Phi\downarrow T_0\colon G \to \mathrm{ALG}(T_0)$$

where T_0 denotes the subtheory of T with $S(T_0) = S(T)$, $\Omega(T_0) = \emptyset$ and $\mathfrak{A}(T_0) = \emptyset$, i.e. $\mathrm{ALG}(T_0)$ is the class of all $S(T)$-sets. An immediate consequence of this fact is the following preservation property of theory-morphisms.

Corollary 4.1.5. *Let* $\varphi\colon T_1 \to T_2$ *be any theory-morphism,* $\Phi\colon G \to \mathrm{ALG}(T_2)$ *be a chain of* T_2*-algebras, and*

$$\big(f(n)\colon n\Phi \to L \mid n \in G_{\text{Nodes}}\big)$$

be a colimit of the diagram Φ *in* $\mathrm{ALG}(T_2)$. *Then*

$$\big(f(n)\colon n(\Phi\varphi) \to L\varphi \mid n \in G_{\text{Nodes}}\big)$$

is a colimit of the diagram

$$\Phi\varphi\colon G \to \mathrm{ALG}(T_1)$$

in $\mathrm{ALG}(T_1)$. ■

Without any additional problems Corollary 4.1.5 can be proved for so-called *directed diagrams* $\Phi\colon G \to \mathrm{ALG}(T_2)$, where a diagram $\Phi\colon G \to \mathrm{ALG}(T_2)$ is called *directed if*

(1) $x\,\mathrm{begin}^G = y\,\mathrm{begin}^G$ and $x\,\mathrm{end}^G = y\,\mathrm{end}^G$ implies $x = y$ for all $x, y \in G_{\text{Edges}}$;

(2) If $x\mathrm{begin}^G = y\mathrm{begin}^G$ for $x, y \in G_{\text{Edges}}$ then there are $x^{\#}, y^{\#} \in G_{\text{Edges}}$ with $x\mathrm{end}^G = x^{\#}\mathrm{begin}^G$, $y\mathrm{end}^G = y^{\#}\mathrm{begin}^G$ and $x^{\#}\mathrm{end}^G = y^{\#}\mathrm{end}^G$;

(3) If $x\mathrm{end}^G = y\mathrm{begin}^G$ for $x, y \in G_{\text{Edges}}$ then there is and edge $z \in G_{\text{Edges}}$ with $x\mathrm{begin}^G = z\mathrm{begin}^G$ and $y\,\mathrm{end}^G = z\mathrm{end}^G$.

A combination of Theorem 4.1.4 with this extended version of Corollary 4.1.5 yields the following closure property.

Theorem 4.1.6. *Any canon model class* $\mathrm{ALG}(T, \Delta) \subseteq \mathrm{ALG}(T)$ *is closed in* $\mathrm{ALG}(T)$ *with respect to colimits of directed diagram in* $\mathrm{ALG}(T, \Delta)$.

Finally we will compare the model classes of canons with the model classes of *hierarchical types* introduced by Broy and Wirsing (1982). A hierarchical type $T = (\Sigma, \mathfrak{A}, P)$ is given by a signature Σ, by a set $\mathfrak{A}$ of axioms, and by a hierarchical type $P = (\Sigma', \mathfrak{A}', P')$, designated as a primitive type of T, so that $\Sigma' \subseteq \Sigma$, $\mathfrak{A}' \subseteq \mathfrak{A}$ and $(\Sigma', \mathfrak{A}') \subset (\Sigma, \mathfrak{A})$. Because of this latter condition, a hierarchical type has only a finite chain of primitive subtypes, the last one, $(\Sigma'', \mathfrak{A}'', \emptyset)$ say, is identified with the non-hierarchical type $(\Sigma'', \mathfrak{A}'')$. In our discussion we will deal for simplicity only with types $(\Sigma, \mathfrak{A}, P)$ with a non-hierarchical primitive subtype $P = (\Sigma', \mathfrak{A}')$. The main difference results from the requirement for model classes of hierarchical types that for a hierarchical $(\Sigma, \mathfrak{A}, P)$-algebra A it is required besides others that $A\downarrow\Sigma'$ is a *finitely generated* partial heterogeneous Σ'-algebra, which in other words means that $A\downarrow\Sigma'$ has no proper subalgebras. It seems that this concept does not allow specifications of polymorphic types; see for instance the specification FINITE PATHS in Section 1.2. Since directed graphs have no constant operations, a directed graph G is finitely generated if it is empty, i.e. if $G_{\text{Nodes}} = G_{\text{Edges}} = \emptyset$.

In the canon specification approach constructors and finitely generated algebras start to play a role if parameter types are instantiated by actual ADTs. At least from this point of view both approaches seem to be rather incomparable. Hierarchical data types seem to be aimed at stepwise specifications of abstract data types in a bottom-up manner and not at polymorphic data types and at specifications of polymorphic functions. The canon specification approach can be used both in bottom-up manner and in top-down manner (see discussion on Definition 4.3.2 and Theorem 4.3.3). The modification-operation on specifications has been introduced especially for using the canon specification approach for top-down design. The formal semantics of this operation is given by pushouts of canon morphisms (see Theorem 4.3.3). It is not known if an analogous theorem can be proved for hierarchical data types.

Another important difference between canons and hierarchical types is that the models of hierarchical types need not satisfy any initiality conditions, but they have to be extensionally equivalent to the initial algebra. In this respect hierarchical types are comparable with canons of behaviour in Chapter 5. Our concept of behaviourally equivalent partial algebras coincides with the extensional equivalence of Broy and Wirsing in the case of finitely generated partial algebras.

If one refrains from partiality, canons and data theories as used in CLEAR, see Burstall, Goguen (1980), the only difference is the fact that a CLEAR parameterized theory has a fixed subtheory representing the parameter, whereas the modification operation of canons allows the modification of an arbitrary subcanon.

The concept of initial restrictions has been generalized in Ehrig, Wagner, Thatcher (1983) to more general data constraints. Based on this generalization in WAGNER, EHRIG (1986) the fundamentals of a general systematic discussion of constraints are developed. It turns out that initial restrictions, which are called linear constraints by WAGNER and EHRIG, are sufficient for practical requirements.

4.2. Computability and definability on parameterized abstract data types

In Chapter 1 the usefulness of functional enrichments of specifications is demonstrated by several examples. In this section we investigate the question of wether a new data type or a new operation can be reduced effectively to the given data types and given operations.

For the sake of conceptual simplicity a specification should distinguish explicitly a minimal set of fundamental operations and should characterize the other ones as computable by derivation from the fundamental operations. This requires a mathematically precise notion of a computable extension of a specification. As announced earlier we prove in this section that the concept of structural induction on partial algebras, or in other words the concept of initial restrictions in algebraic theories with partial operations, is sufficient for the definition of computable extensions of specifications. This purely algebraic version of parameterized algorithms was suggested by H. KAPHENGST in KAPHENGST and REICHEL (1971) and improved in KAPHENGST (1981). The concept of a definable extension of a specification was introduced by K. BENECKE in his thesis in which he also proved some necessary conditions for computable and uniquely specifiable extensions, Theorem 4.2.4 and Theorem 4.2.8 below.

Roughly speaking, a parameterized algorithm $\mathfrak{a}$ associates uniformly with every algebra $A \in \mathrm{ALG}(T, \Delta)$ of a canon model class a function $A\mathfrak{a}$ in such a way that $A\mathfrak{a}$ is a partial recursive function provided A is a computable algebra, i.e. the carrier sets A_s, $s \in S(T)$, are recursively enumerable sets and the fundamental operations σ^A, $\sigma \in \Omega(T)$, are partial recursive functions. Analogously, a parameterized computable set construction $\mathfrak{r}$ associates uniformly with every algebra $A \in \mathrm{ALG}(T, \Delta)$ a recursively enumerable set $A\mathfrak{r}$ provided A is a computable algebra.

Since the parameterization is always represented by a class of models of a canon (T, Δ) we use the notions (T, Δ)-algorithm and (T, Δ)-rest (recursively enumerable set).

Definition 4.2.1. Let (T, Δ) be any canon. A correspondence $\mathfrak{r}$ that associates with every algebra $A \in \mathrm{ALG}(T, \Delta)$ a set $A\mathfrak{r}$ is called a (T, Δ)-*reset* if there is a finite algebraic theory $T^{\#}$ including the algebraic theory T, a finite $S(T)$-sorted system of variables $w\colon X \to S(T)$ and a finite set $(w\colon G)$ of existence equations on $w\colon X \to S(T)$ with operators of $\Omega(T^{\#})$ such that

(1) for every $A \in \mathrm{ALG}(T, \Delta)$ there exists a $T^{\#}$-algebra $A^{\#}$ with $A^{\#}\downarrow T = A$ which is freely generated by A:

(2) $A\mathfrak{r} = A^{\#}_{(w:G)}$ for every $A \in \mathrm{ALG}(T, \Delta)$ where $A^{\#}$ is used in the sense of (1).

$(T^{\#}, G, w)$ is called a *representation* of the (T, Δ)-reset. ■

Definition 4.2.2. Let (T, Δ) be any canon. A correspondence $\mathfrak{a}$ that associates with every algebra $A \in \mathrm{ALG}(T, \Delta)$ a function $A\mathfrak{a}$ is said to be a (T, Δ)-*algorithm,* if there is a finite algebraic theory $T^{\#}$ including the algebraic theory T and an operator $\sigma \in \Omega(T^{\#})$ such that

(1) $\sigma\colon w \to s$ in $\Omega(T^{\#})$ with $w\colon [n] \to S(T)$, $s \in S(T)$;

(2) every $A \in \mathrm{ALG}(T, \Delta)$ generates freely a $T^{\#}$-algebra $A^{\#}$ with $A^{\#}{\downarrow}T = A$;

(3) $A\mathfrak{a} = \sigma^{A^{\#}}$ for every $A \in \mathrm{ALG}(T, \Delta)$ where $A^{\#}$ is used in the sense of (2).

$(T^{\#}, \sigma)$ is called a *representation* of the (T, Δ)-algorithm $\mathfrak{a}$. ■

It is evident that for every (T, Δ)-algorithm $\mathfrak{a}$ represented by $(T^{\#}, \sigma)$ with $\sigma\colon w \to s$ the triple $(T^{\#}, \operatorname{def} \sigma, w)$ is a representation of a (T, Δ)-reset.

Examples of the notions introduced in the preceding definitions can easily be taken from the specifications of Chapter 1.

However, we suppose that these examples do not sufficiently justify the preceding definitions. Therefore we will give a further justification by proving that the introduced concepts coincide with the classical ones in the case $(T, \Delta) = \mathfrak{C}_{\mathrm{NAT}_0}$ with

NAT_0 **is definition**
  sorts N
  oprn $0 \to N$
    $s(N) \to N$
  axioms none
end NAT_0

Theorem 4.2.3. *The* NAT_0*-algorithms are precisely the partial recursive functions on natural numbers, and the* NAT_0*-resets are precisely the recursively enumerable sets of n-tuples of natural numbers.*

Proof. Without loos of generality we may assume that

$$\mathbb{N} = T(\Sigma(\mathrm{NAT}_0), \lambda)_N .$$

First we prove that every NAT_0-reset is recursively enumerable. Therefore let $(T^{\#}, G, w)$ with $w\colon [n] \to \{N\}$ be any representation of a NAT_0-reset. If $A^{\#} \in \mathrm{ALG}(T^{\#})$ is freely generated by $\mathbb{N}$ and $A^{\#}{\downarrow}T(\mathrm{NAT}_0) = \mathbb{N}$, then $A^{\#}$ is isomorphic to $F(\mathfrak{A}(T^{\#}), \emptyset, \lambda)$ according to the proof of Theorem 3.5.6 and Corollary 3.5.7. According to Theorem 3.2.3

$$\boldsymbol{t} \in \mathbb{N}_w = A^{\#}_w \subseteq T(\Sigma(T^{\#}), \lambda)_w$$

is a solution of $(w\colon G)$ in $A^\#$ if and only if

$$[\boldsymbol{t}, \emptyset] \in F\big(\mathfrak{A}(T^\#), \emptyset, \lambda\big)_{(w:G)},$$

i.e. iff

$$\mathfrak{A}(T^\#) \models (\lambda\colon \emptyset \to G\tilde{\boldsymbol{t}}).$$

The results of Section 3.1 imply that the set of all finite sets of existence equations $(\lambda\colon H)$, with operators taken from $\Omega(T^\#)$, satisfying

$$\mathfrak{A}(T^\#) \models (\lambda\colon \emptyset \to H)$$

is recursively enumerable.

The function that associates with every term $\boldsymbol{t} \in \mathbb{N}_w$ the finite set of existence equations $(\lambda\colon G\tilde{\boldsymbol{t}})$ is evidently a computable function, and so $A^\#_{(w:G)}$ is representable as the inverse image of a recursively enumerable set with respect to the computable function just described. This proves the recursive enumerability of $A^\#_{(w:G)}$.

Next we prove that every NAT_0-algorithm is a computable function. Let $(T^\#, \sigma)$ be any representation of a NAT_0-algorithm. As above we use the $T^\#$-algebra $F\big(\mathfrak{A}(T^\#), \emptyset, \lambda\big)$ and apply the results of Section 3.1. As a consequence we can state that the set

$$R = \big\{(t_1, t_2) \in T\big(\Sigma(T^\#), \lambda\big) \times T\big(\Sigma(\mathrm{NAT}_0), \lambda\big) \mid \mathfrak{A}(T^\#) \models (\lambda\colon \emptyset \to t_1 \overset{\mathrm{e}}{=} t_2)\big\}$$

is recursively enumerable. Because of

$$F\big(\mathfrak{A}(T^\#), \emptyset, \lambda\big)\!\downarrow T(\mathrm{NAT}_0) \leftrightarrows T\big(\Sigma(\mathrm{NAT}_0), \lambda\big) = \mathbb{N}$$

this set R is the graph of a partial recursive function f_R. If $\boldsymbol{t} \in \operatorname{dom} \sigma^{A^\#}$ then we obtain

$$\mathfrak{A}(T^\#) \models (\lambda\colon \emptyset \to \boldsymbol{t}\sigma^{T(\Sigma(T^\#),\lambda)} \overset{\mathrm{e}}{=} \boldsymbol{t}\sigma^{T(\Sigma(T^\#),\lambda)})$$

and

$$\boldsymbol{t}\sigma^{A^\#} = (\boldsymbol{t}\sigma^{T(\Sigma(T^\#),\lambda)})\, f_R \in T\big(\Sigma(\mathrm{NAT}_0), \lambda\big).$$

According to the first part of this proof $\sigma^{A^\#}$ is the restriction of the composition

$$\sigma^{T(\Sigma(T^\#),\lambda)} \circ f_R$$

of two computable functions to the recursively enumerable set

$$\operatorname{dom} \sigma^{A^\#}$$

and so $\sigma^{A^\#}$ itself is partial recursive.

So far we have proved the compatibility of the introduced concepts with the classical ones. It remains to prove that they are strong enough. We complete the proof by showing that the recursive functionals of superposition, primitive recursion and minimization, which generate the set of partial recursive functions starting with the constant zero, the successor function and the projections, can be expressed by the introduced concepts.

By Definition 4.2.2 it is evident that the constant function zero, the successor function and the projections are NAT_0-algorithms, and that the superposition of NAT_0-algorithms is also a NAT_0-algorithm.

Due to the Theorem of KLEENE, see ROGERS Jr. (1967), it suffices to show that for all total NAT_0-algorithms

$$g\colon \mathbb{N}^k \to \mathbb{N}, \qquad h\colon \mathbb{N}^{k+2} \to \mathbb{N}$$

the function

$$r(g, h)\colon \mathbb{N}^{k+1} \to \mathbb{N}$$

defined by the functional of primitive recursion is a NAT_0-algorithm, and that for every total NAT_0-algorithm

$$g\colon \mathbb{N}^{k+1} \to \mathbb{N}$$

the partial function

$$\min_y\big((x_0, \ldots, x_{k-1}, y)\, g = 0\big)$$

defined by minimization, or by the μ-operator, is also a NAT_0-algorithm.

Let $(T_1^{\#}, \sigma_1)$, $(T_2^{\#}, \sigma_2)$ be representations of $g\colon \mathbb{N}^k \to \mathbb{N}$ and $h\colon \mathbb{N}^{k+2} \to \mathbb{N}$, respectively. Then there exists a finite extension $T^{\#}$ of $T(\mathrm{NAT}_0)$ containing σ_1 and σ_2 in such a way that $(T^{\#}, \sigma_1)$, $(T^{\#}, \sigma_2)$ is a representation of the function g, h, respectively. Now we set

$T^{\#\#}$ is $T^{\#}$ with
  oprn $r(N^{k+1}) \to N$
  axioms $x_0, x_1, \ldots, x_k\colon N$
    $r(x_0, x_1, \ldots, x_{k-1}, 0) = \sigma_1(x_0, \ldots, x_{k-1})$
    $r(x_0, x_1, \ldots, x_{k-1}, s(x_k)) = \sigma_2(x_0, \ldots, x_{k-1}, r(x_0, \ldots, x_k), x_k)$

and we obtain a representation $(T^{\#\#}, r)$ of $r(g, h)$.

Now let $(T^{\#}, \varrho)$ be a representation of a NAT_0-algorithm $g\colon \mathbb{N}^{k+1} \to \mathbb{N}$. For the specification of the μ-operator we need some auxiliary NAT_0-algorithms. For that reason we begin with

$T_1^{\#}$ is $T^{\#}$ with
  oprn $l(N,N) \to N$
    $\mathrm{eq}(N,N) \to N$
  axioms $x, x_1, x_2\colon N$
    $l(0,x) = 0$
    if $l(x_1,x_2) = 0$ **then** $l(s(x_1),s(x_2)) = 0$
    $l(s(x),0) = s(0)$
    if $l(x_1,x_2) = s(0)$ **then** $l(s(x_1),s(x_2)) = s(0)$
    if $l(x_1,x_2) = 0,\ l(x_2,x_1) = 0$ **then** $\mathrm{eq}(x_1,x_2) = 0$
    if $l(x_1,x_2) = s(0)$ **then** $\mathrm{eq}(x_1,x_2) = s(0)$
    if $\mathrm{eq}(x_1,x_2) = s(0)$ **then** $\mathrm{eq}(x_2,x_1) = s(0)$

$(T_1^\#, l)$, $(T_1^\#, \mathrm{eq})$ are representations of the NAT_0-algorithms $l': \mathbb{N} \times \mathbb{N} \to \mathbb{N}$, $\mathrm{eq}': \mathbb{N} \times \mathbb{N} \to \mathbb{N}$, respectively, with

$$l'(x, y) = 0 \quad \text{iff} \quad x \leq y \quad \text{for all} \quad x, y \in \mathbb{N},$$
$$l'(x, y) = 1 \quad \text{iff} \quad x > y \quad \text{for all} \quad x, y \in \mathbb{N},$$
$$\mathrm{eq}'(x, y) = 0 \quad \text{iff} \quad x = y \quad \text{for all} \quad x, y \in \mathbb{N},$$
$$\mathrm{eq}'(x, y) = 1 \quad \text{iff} \quad x \neq y \quad \text{for all} \quad x, y \in \mathbb{N}.$$

In the next step we set

$T_2^\#$ is $T_1^\#$ with
  sorts P
  oprn $e \to P$
    $v(N,P) \to P$
    $\min(P) \to N$
  axioms $x,x_1,x_2: N,\ p: P$
    $v(x_1,v(x_2,p)) = v(x_2,v(x_1,p))$
    $v(x,v(x,p)) = v(x,p)$
    $\min(e) = 0$
    if $l(x,\min(p)) = s(0)$ **then** $\min(v(x,p)) = x$
    if $l(x,\min(p)) = s(0)$ **then** $\min(v(x,p)) = \min(p)$

If $A^\# \in \mathrm{ALG}(T_2^\#)$ is freely generated by $\mathbb{N}$, then $A^\# \downarrow T(\mathrm{NAT}_0) = \mathbb{N}$ and $A_P^\#$ is up to isomorphism the set of all finite subsets of $\mathbb{N}$ with $e^{A^\#} = \emptyset$ and

$$v^{A^\#}(x, p) = p \cup \{x\}$$

for every $p \in A_P^\#$, $x \in \mathbb{N}$, and

$$\min{}^{A^\#}: A_P^\# \to \mathbb{N}$$

associates with every finite set of natural numbers its minimal element with respect to the natural ordering.

The specification of the μ-operator is finally given by

$T_3^\#$ is $T_2^\#$ with
  sorts Q
  oprn $q_0 \to Q$
    $q(N^{k+1}) \to Q$
    $d(N^{k+1}) \to P$
    $m(x_0{:}N, \ldots, x_{k-1}{:}N$ **iff** $q(x_0, \ldots, x_{k-1},0)=q_0) \to N$
  axioms $x_0,x_1, \ldots, x_k,x_{k+1}: N$
    if $\varrho(x_0, \ldots, x_{k-1},x_k)=0,\ l(x_{k+1},x_k) = 0$
      then $q(x_0, \ldots, x_{k-1},x_{k+1}) = q_0$ **fi**
    if $\varrho(x_0, \ldots, x_{k-1},0)=0$ **then** $d(x_0, \ldots, x_{k-1},0)=v(0,e)$
    if $\mathrm{eq}(\varrho(x_0, \ldots, x_{k-1},0),0)=s(0)$ **then** $d(x_0, \ldots, x_{k-1},0)=e$

if $\varrho(x_0, \ldots, x_{k-1}, s(x_k)) = s(0)$ **then**
$\quad d(x_0, \ldots, x_{k-1}, s(x_k)) = v(s(x_k), d(x_0, \ldots, x_k))$ **fi**
if eq $(\varrho(x_0, \ldots, x_{k-1}, s(x_k)), 0) = s(0)$ **then**
$\quad d(x_0, \ldots, x_{k-1}, s(x_k)) = d(x_0, \ldots, x_k)$ **fi**
if $\varrho(x_0, \ldots, x_{k-1}, x_k) = 0$ **then**
$\quad m(x_0, \ldots, x_{k-1}) = \min(d(x_0, \ldots, x_k))$ **fi**

Up to isomorphism an algebra $A^{\#} \in \mathrm{ALG}(T_3^{\#})$ freely generated by $\mathbb{N}$ is characterized by

$$A^{\#}\downarrow T(\mathrm{NAT}_0) = \mathbb{N},$$

$$A_P^{\#} = \text{set of all finite subsets of } \mathbb{N},$$

$$A_Q^{\#} = \{q_0\} \cup \{(x_0, \ldots, x_k) \in \mathbb{N}^{k+1} \mid (x_0, \ldots, x_{k-1}, y)\, g = 0 \text{ for all } y \leq x_k\},$$

$$e^{A^{\#}} = \emptyset,$$

$$(x, y)\, v^{A^{\#}} = \{x\} \cup y \quad \text{for} \quad x \in \mathbb{N}, \quad y \in A_P^{\#},$$

$$y \min^{A^{\#}} = \min \{x \in \mathbb{N} \mid x \in y\} \quad \text{for all} \quad y \in A_P^{\#} \quad \text{with} \quad y \neq \emptyset,$$

$$\emptyset \min^{A^{\#}} = 0,$$

$$q_0^{A^{\#}} = q_0,$$

$$(x_0, \ldots, x_k)\, q^{A^{\#}} = \begin{cases} q_0 & \text{if there is at least one } y \in \mathbb{N} \text{ with } x_k \leq y \text{ and} \\ & (x_0, \ldots, x_{k-1}, y)\, g = 0, \\ (x_0, \ldots, x_k) & \text{else,} \end{cases}$$

$$(x_0, \ldots, x_k)\, d^{A^{\#}} = \{x \in \mathbb{N} \mid x \leq x_k, (x_0, \ldots, x_{k-1}, x)\, g = 0\},$$

$$(x_0, \ldots, x_{k-1})\, m^{A^{\#}} \quad \text{is defined if and only if there is a } y \in \mathbb{N} \text{ with}$$
$$(x_0, \ldots, x_{k-1}, y)\, g = 0,$$

$$\text{i.e. iff } (x_0, \ldots, x_{k-1}, 0)\, q^{A^{\#}} = q_0,$$

if $(x_0, \ldots, x_{k-1})\, m^{A^{\#}}$ is defined, then

$(x_0, \ldots, x_{k-1})\, m^{A^{\#}} =$ smallest natural number y with $(x_0, \ldots, x_{k-1}, y)\, g = 0$.

This proves that $(T_3^{\#}, m)$ is a representation of the NAT_0-algorithm defined by the μ-operator

$$\min_y\big((x_0, \ldots, x_{k-1}, y)\, g = 0\big). \qquad ■$$

It was not our aim to give the shortest proof of the justification theorem. A side effect of this proof should be a further illustration of the developed approach. Therefore we split up the specification of the μ-operator into isolated modules.

We will not discuss the justification of the introduced concepts in all details. A more profound result in this direction can be found in Kaphengst (1981).

We resume the problem of deciding if an intended functional enrichment of a specification can be specified by an enrichment by definition or not. A first result in this direction is the following property of (T, Δ)-algorithms and (T, Δ)-resets.

Theorem 4.2.4. *(T, Δ)-algorithms and (T, Δ)-resets are compatible with homomorphisms between (T, Δ)-algebras, i.e. if $\mathfrak{a}$ is any (T, Δ)-algorithm with*

$$A\mathfrak{a}: A_w \multimap A_s, \qquad w: X \to S(T), \qquad s \in S(T)$$

for every $A \in \mathrm{ALG}(T, \Delta)$, and if $\mathfrak{r}$ is any (T, Δ)-reset with

$$A\mathfrak{r} \subseteq A_v, \qquad v: Y \to S(T)$$

for every $A \in \mathrm{ALG}(T, \Delta)$, then the following conditions are satisfied for every homomorphism $f: A \to B$ with $A, B \in \mathrm{ALG}(T, \Delta)$.

(1) $\boldsymbol{a} \in \mathrm{dom}(A\mathfrak{a}) \subseteq A_w$ *implies* $\boldsymbol{a}f_w \in \mathrm{dom}(B\mathfrak{a})$ *and* $\big(\boldsymbol{a}(A\mathfrak{a})\big) f_s = (\boldsymbol{a}f_w)\,(B\mathfrak{a})$;

(2) $\boldsymbol{a} \in A\mathfrak{r} \subseteq A_v$ *implies* $\boldsymbol{a}f_v \in B\mathfrak{r}$.

Proof. The compatibility with homomorphisms between (T, Δ)-algebras is an immediate consequence of the definition. Let $(T^{\#}, \sigma)$ be a representation of $\mathfrak{a}$: then

$$A\mathfrak{a} = \sigma^F, \qquad B\mathfrak{a} = \sigma^{F'}, \qquad A = F{\downarrow}T, \qquad B = F'{\downarrow}T$$

if $F, F' \in \mathrm{ALG}(T^{\#})$ are freely generated by A, B, respectively. Corresponding to the homomorphism $f: A \to B$ there exists exactly one homomorphism $f^{\#}: F \to F'$ with $f^{\#}{\downarrow}T = f$ and so $f_w = f^{\#}_w$ for $w: X \to S(T)$. Sinde $f^{\#}$ is a homomorphism we obtain condition (1) above. Since homomorphisms preserve solutions of sets of existence equations condition (2) can be proved in the same way. ■

Even if Theorem 4.2.4 can be proved easily it seems to be of some practical value, as was demonstrated by its application in Chapter 1.

The concept of canons makes available not only a formalization of computable extensions of specifications. In the same way as first-order theories can be extended by definition canons can be extended.

Definition 4.2.5. Let (T, Δ) be any canon. A correspondence $\mathfrak{a}$ that associates with every $A \in \mathrm{ALG}(T, \Delta)$ a mapping $A\mathfrak{a}$ is called a *definable (T, Δ)-mapping* if there is a finite extension $T^{\#}$ of the algebraic theory T with $\sigma \in \Omega(T^{\#})$ such that

(1) $\sigma: w \to s$, $w: [n] \to S(T)$, $s \in S(T)$;

(2) For every $A \in \mathrm{ALG}(T, \Delta)$ there exists at least one $A^{\#} \in \mathrm{ALG}(T^{\#})$ with $A^{\#}{\downarrow}T = A$;

(3) For all $A^{\#}, B^{\#} \in \mathrm{ALG}(T^{\#})$ with $A^{\#}{\downarrow}T = B^{\#}{\downarrow}T = A \in \mathrm{ALG}(T, \Delta)$ it follows that $\sigma^{A^{\#}} = \sigma^{B^{\#}} = A\mathfrak{a}$.

$(T^{\#}, \sigma)$ is called a *representation* of the definable (T, Δ)-mapping $\mathfrak{a}$.

A correspondence $\mathfrak{r}$ that associates with every $A \in \mathrm{ALG}(T, \Delta)$ a set $A\mathfrak{r}$ is said to be a *definable* (T, Δ)*-set* if there is a finite extension $T^{\#}$ of the algebraic theory T, a finite $S(T)$-sorted system of variables $w\colon X \to S(T)$ and a finite set of existence equations $(w\colon G)$ using operators of $\Omega(T^{\#})$ such that

(1) For every $A \in \mathrm{ALG}(T, \Delta)$ there is at least one $A^{\#} \in \mathrm{ALG}(T, \Delta)$ with $A^{\#}{\downarrow}T = A$;

(2) For all $A^{\#}$, $B^{\#} \in \mathrm{ALG}(T, \Delta)$ with $A^{\#}{\downarrow}T = B^{\#}{\downarrow}T = A \in \mathrm{ALG}(T, \Delta)$ it follows that $A^{\#}_{(w:G)} = B^{\#}_{(w:G)} = A\mathfrak{r}$.

$(T^{\#}, G, w)$ is called a *representation* of the definable (T, Δ)-set $\mathfrak{r}$. ■

These two different kinds of extensions give rise to the study of the relations between the concepts of computability and definability over canons.

By similar ideas as used in the proof of Theorem 4.2.3 one can prove that over the totally restricted canon NAT_0 every NAT_0-algorithm and every NAT_0-reset is a definable NAT_0-mapping and a definable NAT_0-set, respectively. However, in general the two concepts are independent.

Corollary 4.2.6. *Computability and definability over general canons are incomparable concepts.*

Proof. We prove this corollary by giving a (T_1, Δ_1)-reset that is not definable and a definable (T_2, Δ_2)-mapping that is not a (T_2, Δ_2)-algorithm.

Let the canon (T_1, Δ_1) be given by $T_1 = T_{\mathrm{NAT}_0}$ and $\Delta_1 = \emptyset$, so that $\mathrm{ALG}(T_1, \Delta_1) = \mathrm{ALG}(T_{\mathrm{NAT}_0})$. For this very simple algebraic theory we consider the (T_1, Δ_1)-reset $\mathfrak{r}$ that associates with every $A \in \mathrm{ALG}(T_{\mathrm{NAT}_0})$ the set

$$A\mathfrak{r} = \{x \in A_N \mid x = 0^A \underbrace{s^A \ldots s^A}_{n\text{-times}},\ n \in \mathbb{N}\}$$

of all those elements $x \in A_N$ reachable from 0^A by finitely many applications of $s^A\colon A_N \to A_N$. A representation of this (T_1, Δ_1)- reset is given by

$$\left(T_1^{\#}, q(x) = q_0, N\right)$$

with

$T_1^{\#}$ **is** T_{NAT_0} **with**
sorts Q
oprn $q_0 \to Q$
$q(N) \to Q$
axioms $x\colon N$
$q(0) = q_0$
if $q(x) = q_0$ **then** $q(s(x)) = q_0$

If we assume that this $(T_{\mathrm{NAT}_0}, \emptyset)$-reset is definable and that (T_3, G, N) is a representation of it, then for every $A \in \mathrm{ALG}(T_3)$

$$A_{(N:G)} = (A{\downarrow}T_{\mathrm{NAT}_0})\,\mathfrak{r}$$

where $\mathfrak{r}$ denotes the $(T_{\mathrm{NAT}_0}\, \emptyset)$-reset represented by $\big(T_1^{\#}, q(x) = q_0, N\big)$. By adding each existence equation of $(N\colon G)$ as an axiom to the set $\mathfrak{A}(T_3)$ of axioms of T_3 we obtain the theory $T_3^{\#}$. Now for every $B \in \mathrm{ALG}(T_3^{\#})$ it follows that

$$B_N = B_{(N:G)} = (B{\downarrow}T_{\mathrm{NAT}_0})\,\mathfrak{r}$$

and so the set B_N is at most countable for every $B \in \mathrm{ALG}(T_3^{\#})$. However, this property contradicts the following property of every class of models of a theory T: If there is an algebra $A \in \mathrm{ALG}(T)$ where A_s consists of at least two elements for some $s \in S(T)$ then for every cardinality α there exists an algebra $B \in \mathrm{ALG}(T)$ with $\alpha \leq \mathrm{card}\, B_s$. This property can be proved as follows: Let M be a set with card $M = \alpha$ and $w\colon M \to S(T)$ an $S(T)$-sorted system of variables with $xw = s$ for every $x \in M$. Now we consider the T-algebra $F\big(\mathfrak{A}(T), \emptyset, w\big)$ and the universal solution

$$\boldsymbol{e} \in F\big(\mathfrak{A}(T), \emptyset, w\big)_{(w:\emptyset)} = F\big(\mathfrak{A}(T), \emptyset, w\big)_w.$$

The property is proved if we can show that $\boldsymbol{e}$ is an injective assignment. Let $x_1, x_2 \in M$ be any two variables. Then there is an assignment $\boldsymbol{f} \in A_w$ with $x_1\boldsymbol{f} = x_2\boldsymbol{f}$ because card $A_s \geq 2$. This implies the existence of a homomorphism

$$f^{\#}\colon F\big(\mathfrak{A}(T), \emptyset, w\big) \to A$$

with $x_1\boldsymbol{e}f_s^{\#} = x_1\boldsymbol{f}$, $x_2\boldsymbol{e}f_s^{\#} = x_2\boldsymbol{f}$ and so $x_1\boldsymbol{e} \neq x_2\boldsymbol{e}$ because of $x_1\boldsymbol{f} \neq x_2\boldsymbol{f}$.

For the second part of the proof let the canon (T_2, Δ_2) be given by

(T_2, Δ_2) **is definition**
  sorts Bool
  oprn true $\to$ Bool
    false $\to$ Bool
  axioms none
  with requirement
  sorts M
  oprn none
  axioms none
end (T_2, Δ_2).

The identity of elements of sort M is a definable (T_2, Δ_2)-mapping represented by $(T_2^{\#}, \mathrm{id})$ with

$T_2^{\#}$ **is** T_2 **with**
  oprn id$(M,M) \to$ Bool
  axioms $x,y\colon M$
    id(x,x) = true
    if id(x,y) = true **then** $x = y$

By means of Theorem 4.2.4 one can easily prove that this definable (T_2, Δ_2)-mapping is not a (T_2, Δ_2)-algorithm because it is evidently not compatible with every homomorphism between (T_2, Δ_2)-algebras. ■

One should take note of the fact that definability over canons is not explicitly indicated within specifications. The keyword 'definition' does not indicate the specification of a definable (T, Δ)-set or of a definable (T, Δ)-mapping; on the contrary it usually indicates the specification of a (T, Δ)-algorithm or of a (T, Δ)-reset. Definability has to be expressed by enrichments by requirements.

In practical applications an intended (T, Δ)-mapping is very often specified by alternating enrichments by definition and requirement. A typical example for that case is the specification of the predicate on directed graphs indicating the existence of a finite directed path from one node to another one as specified in C-D-GRAPH; see page 38 and 39. This gives rise to the following definition.

Definition 4.2.7. Let (T, Δ) be any canon. A correspondence $\mathfrak{a}$ that associates with every $A \in \mathrm{ALG}(T, \Delta)$ a mapping

$$A\mathfrak{a}\colon \mathrm{dom}(A\mathfrak{a}) \to A_s$$

with $\mathrm{dom}(A\mathfrak{a}) \subseteq A_w$, $w\colon [n] \to S(T)$, $s \in S(T)$, is called a *uniquely specifiable (T, Δ)-mapping* if there is a canon $(T^\#, \Delta^\#)$ with $T \subseteq T^\#$, $\Delta \subseteq \Delta^\#$ and an operator $\sigma \in \Omega(T^\#)$ such that

(1) For every $A \in \mathrm{ALG}(T, \Delta)$ there exists at least one algebra $A^\# \in \mathrm{ALG}(T^\#, \Delta^\#)$ with $A = A^\# \downarrow T$;

(2) For all $A^\#, B^\# \in \mathrm{ALG}(T^\#, \Delta^\#)$ with $A^\# \downarrow T = B^\# \downarrow T = A \in \mathrm{ALG}(T, \Delta)$ it follows $\sigma^{A^\#} = \sigma^{B^\#} = A\mathfrak{a}$.

$(T^\#, \Delta^\#, \sigma)$ is called a *representation* of the uniquely specifiable (T, Δ)-mapping $\mathfrak{a}$.

A correspondence $\mathfrak{r}$ that associates with every $A \in \mathrm{ALG}(T, \Delta)$ a set $A\mathfrak{r} \subseteq A_w$ with $w\colon [n] \to S(T)$, $s \in S(T)$ is called a *uniquely specifiable (T, Δ)-set* if there is a canon $(T^\#, \Delta^\#)$ with $T \subseteq T^\#$, $\Delta \subseteq \Delta^\#$ and a finite set of existence equations $(w\colon G)$ using operators of $\Omega(T^\#)$ such that

(1) For every $A \in \mathrm{ALG}(T, \Delta)$ there exists at least one algebra $A^\# \in \mathrm{ALG}(T^\#, \Delta^\#)$ with $A = A^\# \downarrow T$;

(2) For all $A^\#, B^\# \in \mathrm{ALG}(T^\#, \Delta^\#)$ with $A^\# \downarrow T = B^\# \downarrow T = B^\# \downarrow T = A \in \mathrm{ALG}(T, \Delta)$ it follows $A\mathfrak{r} = A^\#_{(w:G)} = B^\#_{(w:G)}$.

$(T^\#, \Delta^\#, G, w)$ is called a *representation* of the uniquely specifiable (T, Δ)-set $\mathfrak{r}$. ■

By a slight modification of the preceding example of a (T_1, Δ_1)-reset that is not definable, see proof of Corollary 4.2.6, we can give in the following an example of a uniquely specifiable (T, Δ)-mapping that is neither definable nor computable. This example proves that Definition 4.2.7 represents a proper generalization.

The correspondence $\mathfrak{a}$ may associate with every algebra $A \in \mathrm{ALG}(T(\mathrm{NAT}_0), \emptyset)$ the predicate

$$A\mathfrak{a}\colon A_N \to A_{\mathrm{Bool}}$$

with $x(A\mathfrak{a}) = \text{true}$ if x is derivable from 0^A by a finite number of applications of $\sigma^A\colon A_N \to A_N$ and $x(A\mathfrak{a}) = \text{false}$ otherwise. The specification of this predicate is given by the following canon DERIVABLE:

DERIVABLE is NAT_0 **with definition**
  sorts Q
  oprn $q_0 \to Q$
    $q(N) \to Q$
  axioms $x\colon N$
    $q(0) = q_0$
    if $q(x) = q_0$ **then** $q(s(x)) = q_0$
  with requirement
  oprn $p(N) \to \mathrm{Bool}$
  axioms $p(N) \to \mathrm{Bool}$
  axioms $x\colon N$
    if $q(x) = q_0$ **then** $p(x) = \text{true}$
    if $p(x) = \text{true}$ **then** $q(x) = q_0$
end DERIVABLE

With this specification (DERIVABLE, p) is a representation of the described $(T_{\mathrm{NAT}_0}, \emptyset)$-predicate. Since $\mathrm{ALG}(T_{\mathrm{NAT}_0}, \emptyset)$ is a model class of a first-order theory similar considerations to those above enable us to see that this $(T_{\mathrm{NAT}_0}, \emptyset)$-predicate is not a definable $(T_{\mathrm{NAT}_0}, \emptyset)$-mapping

Theorem 4.2.4 implies that $\mathfrak{a}$ is not a $(T_{\mathrm{NAT}}, \emptyset)$-algorithm provided one can define a homomorphism $f\colon A \to B$ with A and B in $\mathrm{ALG}(T_{\mathrm{NAT}_0}, \emptyset)$ which is not compatible with $\mathfrak{a}$. Such a homomorphism is given by

$$A_N = \mathbb{N},\quad 0^A = 0,\quad x\sigma^A = x + 2 \quad \text{for every} \quad x \in \mathbb{N},$$
$$B_N = \mathbb{N},\quad 0^B = 0,\quad x\sigma^B = x + 1 \quad \text{for every} \quad x \in \mathbb{N},$$
$$(2x)\, f_N = x,\quad (2x+1)\, f_N = x \quad \text{for every} \quad x \in \mathbb{N},$$

and

$$x f_{\mathrm{Bool}} = x \quad \text{for} \quad x \in \{\text{true, false}\}$$

because

$$3(A\mathfrak{a})\, f_{\mathrm{Bool}} = \text{false}, \qquad 3 f_N(B\mathfrak{a}) = 1(B\mathfrak{a}) = \text{true},$$

i.e.

$$3(A\mathfrak{a})\, f_{\mathrm{Bool}} \neq 3 f_N(B\mathfrak{a}).$$

We finish this section with a property of uniquely specifiable (T, Δ)-mappings and uniquely specifiable (T, Δ)-sets

Theorem 4.2.8. *Uniquely specifiable (T, Δ)-mapping and uniquely specifiable (T, Δ)-sets are compatible with isomorphisms between (T, Δ)-algebras.*

Proof. Let $\mathfrak{a}$ be any uniquely specifiable (T, Δ)-mapping with

$$A\mathfrak{a}\colon A_w \circ\!\!\rightarrow A_s, \qquad w\colon [n] \to S(T), \qquad s \in S(T)$$

and let $\mathfrak{r}$ be any uniquely specifiable (T, Δ)-set with

$$A\mathfrak{r} \subseteq A_v, \qquad v\colon [m] \to S(T)$$

for every $A \in \mathrm{ALG}(T, \Delta)$. Additionally let $f\colon A \to B$ be an isomorphism with $A, B \in \mathrm{ALG}(T, \Delta)$. According to Definition 4.2.7 there are representations $(T^{\#}, \Delta^{\#}, \sigma)$ and $(T^{\#}, \Delta^{\#}, G, w)$ of $\mathfrak{a}$ and $\mathfrak{r}$, respectively, so that $A^{\#}, B^{\#} \in \mathrm{ALG}(T^{\#}, \Delta^{\#})$ exist with

$$A = A^{\#}\!\downarrow T, \quad B = B^{\#}\!\downarrow T, \quad A\mathfrak{a} = \sigma^{A^{\#}}, \quad B\mathfrak{a} = \sigma^{B^{\#}},$$

and

$$A\mathfrak{r} = A^{\#}_{(v:G)}, \qquad B\mathfrak{r} = B^{\#}_{(v:G)},$$

respectively. We have to prove

(i) $\boldsymbol{a}(A\mathfrak{a})\, f_s = (\boldsymbol{a} f_w)\,(B\mathfrak{a})$ for every $\boldsymbol{a} \in \operatorname{dom} A\mathfrak{a}$ and
(ii) $\boldsymbol{a} f_v \in B\mathfrak{r}$ for every $\boldsymbol{a} \in A\mathfrak{r} \subseteq A_v$.

If we substitute in $A^{\#}$ for every $a \in A^{\#}_s$, $s \in S(T)$ the element $a f_s \in B_s$ we get a $T^{\#}$-algebra $A^{\#\#}$ and an isomorphism $f^{\#}\colon A^{\#} \to A^{\#\#}$ given by $f^{\#}_s = f_s$ if $s \in S(T)$ and $f^{\#}_s = \mathrm{Id}_{A^{\#}_s}$ otherwise. Because of $A^{\#\#}\!\downarrow T = B$ condition (2) of Definition 4.2.7 implies

$$B\mathfrak{r} = A^{\#\#}_{(v:G)} \quad \text{and} \quad B\boldsymbol{a} = \sigma^{A^{\#\#}},$$

respectively. Since $f^{\#}\colon A^{\#} \to A^{\#\#}$ is a homomorphism we obtain

$$\boldsymbol{a}(A\mathfrak{a})\, f_s = \big(\boldsymbol{a}(A\mathfrak{a})\big)\, f^{\#}_s = (\boldsymbol{a}\sigma^{A^{\#}})\, f_s = (\boldsymbol{a} f^{\#}_w)\, \sigma^{A^{\#\#}} = (\boldsymbol{a} f^{\#}_w)\,(B\mathfrak{a}) = (\boldsymbol{a} f_w)\,(B\mathfrak{a})$$

or every $\boldsymbol{a} \in \operatorname{dom}(A\mathfrak{a}) \subseteq A_w$, and

$$\boldsymbol{a} f_v = \boldsymbol{a} f^{\#}_v \in A^{\#\#}_{(v:G)} = B\mathfrak{r}$$

if $\boldsymbol{a} \in A\mathfrak{r} = A^{\#}_{(v:G)}$. ■

4.3. Canon morphisms

Each canon represents intuitively a certain system of concepts including semantically fixed concepts, represented by sorts and operators specified in enrichments by definitions of the empty canon, parameterized concepts, specified in enrichments by definitions, and open fundamental concepts, specified in enrichments by requirement. One advantage of axiomatic specifications is the possibility of minimizing the conceptual distance between the application area and the formal specification of the intended class of problems. In this way an axiomatic specification makes explicitly visible the conceptual distance between the conceptual level of a certain problem and the conceptual level of a given programming language. In most cases the high complexity of a computer program results from this conceptual distance. Therefore the clarity of a computer program can be improved considerably if the representation of the solving algorithm in terms of the conceptual level of the problem and the reduction of the problem level to the level of the programming language are described independently. These two independent dimensions of the complexity of a computer program are very clearly and very well considered in the system CAT, a system for the structured elaboration of correct programs from structured specifications, in Goguen and Burstall (1980).

A further topic that motivates a mathematically precise description of relations between different conceptual levels, i.e. between canons, is the actualization and the spezialization of parameterized spezifications. By actualization we understand the substitution of open concepts of a canon by semantically fixed concepts of another canon, and by specialization we understand the substitution of open concepts of a canon by open concepts of another canon with a smaller range of interpretations.

A mathematical representation of the relations between different conceptual levels is the following notion of canon morphisms.

Definition 4.3.1. Let (T_1, Δ_1), (T_2, Δ_2) be canons. A theory-morphism $\varphi\colon T_1 \to T_2$ (see Definition 3.5.2) is called a *canon morphism* from (T_1, Δ_1) to (T_2, Δ_2), denoted by

$$\varphi\colon (T_1, \Delta_1) \to (T_2, \Delta_2),$$

if for every $A \in \mathrm{ALG}(T_2, \Delta_2)$, $A\varphi \in \mathrm{ALG}(T_1, \Delta_1)$ holds.

A canon (T_1, Δ_1) is called *subcanon* of the canon (T_2, Δ_2) or (T_2, Δ_2) an *enlargement* of (T_1, Δ_1) if T_1 is a subtheory of T_2 and if the inclusion of T_1 in T_2 forms a canon morphism. If (T_1, Δ_1) is a subcanon of (T_2, Δ_2) we write

$$(T_1, \Delta_1) \subseteq (T_2, \Delta_2)$$

for short.

A canon morphism $\varphi\colon (T_1, \Delta_1) \to (T_2, \Delta_2)$ is said to be *persistent* if

(1) for every $B \in \mathrm{ALG}(T_1, \Delta_1)$ there is an algebra $A \in \mathrm{ALG}(T_2, \Delta_2)$ freely generated by B according $\varphi\colon T_1 \to T_2$ with $B = A\varphi$.

(2) for every $A \in \mathrm{ALG}(T_2, \Delta_2)$, A is freely generated by $A\varphi$ according $\varphi\colon T_1 \to T_2$. ■

The introduced notion of a subcanon is a rather weak concept. If we assume that $\mathbb{C} = (T, \Delta)$ is any canon and T_0 any subtheory of T, then

$(T_0, \emptyset)$ would be a subcanon of (T, Δ). But, in this subcanon none of the initial restrictions in Δ have been transferred. We say that an initial restriction $(T_1, \varphi\colon T_2 \to T)$ induces the initial restriction

$$(T_{0,1}, \varphi_0\colon T_{0,2} \to T_0)$$

in the subtheory $T_0 \subseteq T$ if in the following diagram of theory-morphisms both squares I and II are pullbacks (i.e. constructions of inverse images):

$$\begin{array}{ccccc}
T_1 & \xrightarrow{\;\subseteq\;} & T_2 & \xrightarrow{\;\varphi\;} & T \\
\uparrow{\scriptstyle\subseteq} & I & \uparrow{\scriptstyle\subseteq} & II & \uparrow{\scriptstyle\subseteq} \\
T_{0,1} & \xrightarrow{\;\subseteq\;} & T_{0,2} & \xrightarrow{\;\varphi_0\;} & T_0
\end{array}$$

The subcanon (T_0, Δ_0) is said to be *induced by the subtheory* $T_0 \subseteq T$ if

$$\begin{aligned}\Delta_0 = \{(T_{0,1}, \varphi_0\colon T_{0,2} \to T_0) \mid\ & \text{induced in } T_0 \text{ by some initial restriction in } \Delta, \\ & \text{and for every } A \in \mathrm{ALG}(T, \Delta) \\ & A{\downarrow}T_0 \models (T_{0,1}, \varphi_0\colon T_{0,2} \to T_0)\}.\end{aligned}$$

In the case that a canon morphism $\varphi\colon (T_1, \Delta_1) \to (T_2, \Delta_2)$ represents a reduction of a system of concepts described by (T_1, Δ_1) to a system of concepts of a lower level described by (T_2, Δ_2) one can notice that the concept of a canon morphism is not yet adequate. The notion of a reduction includes that a fundamental concept of (T_1, Δ_1) is reduced in general to a composition of concepts that can effectively be built up from the fundamental concepts of (T_2, Δ_2). Unfortunately not every sort name or operator introduced by an enrichment by definition respresents a compounded concept that can be built up effectively from the fundamental concepts. The concepts of (T, Δ)-resets and (T, Δ)-algorithms as introduced in Section 4.2 are a first mathematical approximation of computability on canons. But, these versions do not fit together very well with the notion of a canon morphism since canon morphisms map sort names to sort names and not to sets of existence equations. There are several ways of reconciling these two concepts. One way is given by the following definition.

We start with the theoretical background of the definition. In the previous section the notion of a definable (T, Δ)-mapping was introduced and one could see that for every sort name $s \in S(T)$ the identity relation is definable via a standard enrichment by requirement (see proof of Corollary 4.2.6). However, we know that it is not true to say that for every sort name which is introduced by a definition enrichment the identity relation of that new data type can be specified again by a definition enrichment. According to the results of Chapter 3 and Section 4.2 we know that in general the semantical equivalence of terms, producing values of a sort name introduced by a definition enrichment, is only a recursively enumerable relation in the set of terms. This relation is recursively decidable if and only if there is a recursively enumerable set of representatives of the semantical equivalence, or in other words, if the specified data type has a computable prototype. Since every partial recursive (polymorphic) function can be specified by an enrichment by definition, we can characterize those (polymorphic) data types that have a computable prototype by the statement that the identity relation can be specified by a definition enrichment. If this data type is proper polymorphic, then naturally the identity relations of the parameter types are necessary for that specification.

To make this idea more precise we introduce the following notations. $\mathbb{C} = (T, \Delta)$ may be any canon and $S_0 \subseteq S(T)$ any subset of the set of sort names of $\mathbb{C}$; then

$$\mathbb{C}^{=}(S_0)$$

denotes the canon that is the result of an enrichment by requirement which defines the identity relations

$$s\text{-eq}: ss \rightarrow \text{Bool}$$

for each sort name $s \in S_0$ in the standard way, if the identity relation is not yet contained in $\mathbb{C}$. If $\mathbb{C}$ does not include the specification of the truth values, i.e. if $\mathbb{C}_{\text{BOOL}}$ is not contained in $\mathbb{C}$, then the construction of $\mathbb{C}^{=}(S_0)$ includes as first step an enrichment by definition of $\mathbb{C}$ specifying the truth values. Hence it holds $\mathbb{C}^{=}(\emptyset) = \mathbb{C}$ if $\mathbb{C}_{\text{BOOL}} \subseteq \mathbb{C}$ and $\mathbb{C}^{=}(\emptyset) = \mathbb{C}$ **with** $\mathbb{C}_{\text{BOOL}}$ else. The canon $\mathbb{C}^{=}(S_0)$ is said to be the *enrichment of* $\mathbb{C}$ *by* S_0*-identies.*

Definition 4.3.2. $\mathbb{C}_0 = (T_0, \Delta_0)$ may be a subcanon of $\mathbb{C} = (T, \Delta)$. A sort name s of $\mathbb{C}$ i.e. $s \in S(T)$, is called *recursive over* $\mathbb{C}_0$, if there is a subset $S' \subseteq S(T_0)$ and a persistent enlargement $\mathbb{C}_0^{=}(S') \subseteq \mathbb{C}_1 = (T_1, \Delta_1)$ containing a sort name $s^* \in S(T_1)$ and an operator $s^*\text{-eq}: s^*s^* \rightarrow \text{Bool}$ so that for every $\mathbb{C}$-algebra A there is a $\mathbb{C}_1$-algebra B with $A{\downarrow}T_0 = B{\downarrow}T_0$, $A_s = B^*_s$, and $(s^*\text{-eq})^B: A_s \times A_s \rightarrow \{\text{true,false}\}$ is the identity relation.

An operator $\sigma: s_1 \ldots s_n \rightarrow s_0$ of $\mathbb{C}$ is said to be *recursive over* $\mathbb{C}_0$, if $s_0, s_1, \ldots, s_n$ are recursive over $\mathbb{C}_0$ and if there is a subset $S' \subseteq S(T_0)$ and a persistent enlargement $\mathbb{C}_0^{=}(S') \subseteq \mathbb{C}_1 = (T_1, \Delta_1)$ containing an operator $\sigma^*: s_1^* \ldots s_n^* \rightarrow s_0^*$ so that for every $\mathbb{C}$-algebra A there is a $\mathbb{C}_1$-algebra B with $A{\downarrow}T_0 = B{\downarrow}T_0$, $A_{s_i} = B_{s_i^*}$ for $i = 0, 1, \ldots, n$, and $(\sigma^*)^B = \sigma^A$. ■

If we take the specification C-D-GRAPH of Section 1.2 and define the following enrichment

C-D-GRAPH **is** C-D-GRAPH **with** BOOL
    with requirement
    oprn eq(Path,Path) $\rightarrow$ Bool
    axioms u,v: Path
        eq(u,u) = true
        if eq(u,v) = true **then** $u = v$
end C-D-GRAPH

then one can state:

(i) The sort name 'Path' is recursive over $\mathbb{C}_{\text{D-GRAPH}}$, because of the specification EXAMPLE 2.4 which specifies the identity relation of 'Path' by a persistent enlargement of $\mathbb{C}^{=}_{\text{D-GRAPH}}(\{\text{Nodes,Edges}\})$;

(ii) An immediate consequence of (i) is the recursiveness of the operator 'eq' over $\mathbb{C}_{\text{D-GRAPH}}$, although that operator is introduced by a requirement enrichment;

(iii) Due to Theorem 4.2.4 the operator 'connected' (see in Section 1.2 the discussion following the specification D-GRAPH) is not recursive over $\mathbb{C}_{\text{D-GRAPH}}$.

Notice, the notion of an operator recursive over a subcanon $\mathbb{C}_0 \subseteq \mathbb{C}$ is not identical with the notion of a $\mathbb{C}_0$-algorithm introduced in Definition 4.2.2. The identity of both concepts is achieved if we require $S' = \emptyset$ in the definition of recursivenes of an operator over a subcanon. But, in this case it is not true that a sort name is recursive over a subcanon if and only if its identity relation is recursive over that subcanon.

With respect to the more general concept of recursiveness the applicability of Theorem 4.2.4 is restricted to one-to-one homomorphisms between algebras of the parameter class.

An interesting case is the recursiveness over the empty subcanon. Sort names and operators recursive over the empty subcanon could be called *absolutely recursive*. All known examples of ADTs such as truth values, natural numbers, integers, integer lists, and so on are absolutely recursive and all computable functions on these ADTs correspond to absolutely recursive operator names in initial canons. Analogously, one is only interested in specifications of polymorphic data types that are recursive over the subcanon specifying the parameters. However, one can not say that only the specifications of interest are those which contain only operators recursive over a parameter canon. The first specification of an intended algorithm might very well be given by an operator which is not recursive over the parameter canon $\mathbb{C}_0$, but which is a definable $\mathbb{C}_0$-mapping, see Defintion 4.2.5 and specification C-D-GRAPH. Here we see that enrichments by requirements are not only aimed at specifications of parameter classes.

It is easy to prove that a sort name or an operator which is recursive over a subcanon $\mathbb{C}_0$, whose sort names and operators are all absolutely recursive, is itself absolutely recursive. This fact makes possible specifications of ADTs in a bottom-up manner.

The theoretical justification of the top-down design of ADTs, using polymorphic data types and instantiations of parameters via the modification operation on specifications, is given by the following theorems.

Theorem 4.3.3. *For every pair* $\varepsilon\colon (T_0, \Delta_0) \to (T_1, \Delta_1)$, $\varphi\colon (T_0, \Delta_0) \to (T_2, \Delta_2)$ *of canon morphisms there exists a pushout*

$$\begin{array}{ccc} (T_0, \Delta_0) & \xrightarrow{\ \varepsilon\ } & (T_1, \Delta_1) \\ {\scriptstyle\varphi}\downarrow & & \downarrow{\scriptstyle\varphi^{\#}} \\ (T_2, \Delta_2) & \xrightarrow{\ \varepsilon^{\#}\ } & (T, \Delta) \end{array}$$

Proof. We recall that a canon morphism $\varphi\colon (T_0, \Delta_0) \to (T_2, \Delta_2)$ consists of a pair of functions

$$\varphi_S\colon S(T_0) \to S(T_2), \qquad \varphi_\Omega\colon \Omega(T_0) \to \Omega(T_2).$$

We start with the following constructions

$$\begin{array}{ccc} S(T_0) & \xrightarrow{\ \varepsilon_S\ } & S(T_1) \\ {\scriptstyle\varphi_S}\downarrow & & \downarrow{\scriptstyle\varphi_S^{\#}} \\ S(T_2) & \xrightarrow{\ \varepsilon_S^{\#}\ } & S(T) \end{array} \qquad \begin{array}{ccc} \Omega(T_0) & \xrightarrow{\ \varepsilon_\Omega\ } & \Omega(T_1) \\ {\scriptstyle\varphi_\Omega}\downarrow & & \downarrow{\scriptstyle\varphi_\Omega^{\#}} \\ \Omega(T_2) & \xrightarrow{\ \varepsilon_\Omega^{\#}\ } & \Omega(T) \end{array}$$

with

$$S(T) = \big(S(T_1) + S(T_2)\big)/\varrho_S,$$

$$\Omega(T) = \big(\Omega(T_1) + \Omega(T_2)\big)/\varrho_\Omega$$

where $S(T_1) + S(T_2)$, $\Omega(T_1) + \Omega(T_2)$ are disjoint unions, ϱ_S is the equivalence relation in $S(T_1) + S(T_2)$ generated by the set of pairs

$$\{(s\varepsilon_S, s\varepsilon_S) \mid s \in S(T_0)\}$$

and ϱ_Ω is the equivalence relation in $\Omega(T_1) + \Omega(T_2)$ generated by the set of pairs

$$\{(\sigma\varepsilon_\Omega, \sigma\varepsilon_\Omega) \mid \sigma \in \Omega(T_0)\}.$$

The functions $\varepsilon_S^{\#}$, $\varphi_S^{\#}$, $\varphi_\Omega^{\#}$, $\varphi_\Omega^{\#}$ are defined by

$$s\varepsilon_S^{\#} = [s]_{\varrho_S} \quad \text{for every} \quad s \in S(T_2),$$

$$s\varphi_S^{\#} = [s]_{\varrho_S} \quad \text{for every} \quad s \in S(T_1),$$

$$\sigma\varepsilon_\Omega^{\#} = [\sigma]_{\varrho_\Omega} \quad \text{for every} \quad \sigma \in \Omega(T_2),$$

$$\sigma\varphi_\Omega^{\#} = [\sigma]_{\varrho_\Omega} \quad \text{for every} \quad \sigma \in \Omega(T_1).$$

It is easy to see that

$$\varphi_S \circ \varepsilon_S^{\#} = \varepsilon_S \circ \varphi_S^{\#} \quad \text{and} \quad \varphi_\Omega \circ \varepsilon_\Omega^{\#} = \varepsilon_\Omega \circ \varphi_\Omega^{\#}.$$

Now we define two functions

$$\psi_S: S(T_0) \to S(T), \qquad \psi_\Omega: \Omega(T_1) \to \Omega(T)$$

by

$$\psi_S = \varphi_S \circ \varepsilon_S^{\#} = \varepsilon_S \circ \varphi_S^{\#},$$

Next we define the arity function and the domain conditions for the operators of $\Omega(T)$.

If $\sigma \in \Omega(T_1)$ and there is an operator $\sigma' \in \Omega(T_0)$ with $\sigma': w \to s$ and $\sigma\varphi_\Omega^{\#} = \sigma'\varepsilon_\Omega\varphi_\Omega^{\#}$ then we set

$$\sigma\varphi_\Omega^{\#}: w \circ \psi_S \to s\varphi_S \quad \text{and} \quad \mathrm{def}_T(\sigma\varphi_\Omega^{\#}) = \big(\mathrm{def}_{T_0}(\sigma')\big)\,\psi.$$

Analogously we define the arity and the domain condition if $\sigma \in \Omega(T_2)$, $\sigma'' \in \Omega(T_0)$ with $\sigma\varepsilon_\Omega^{\#} = \sigma''\varphi_\Omega\varepsilon_\Omega^{\#}$.

If $\sigma \in \Omega(T_1)$ with $\sigma: w \to s$ and there is no $\sigma' \in \Omega(T_0)$ with $\sigma\varphi_\Omega^{\#} = \sigma'\varepsilon_\Omega\varphi_\Omega^{\#}$ then we define

$$\sigma\varphi_\Omega^{\#} = w \circ \varphi_S^{\#} \to s\varphi_s \quad \text{and} \quad \mathrm{def}_T(\sigma\varphi_\Omega^{\#}) = \big(\mathrm{def}_{T_1}(\sigma)\big)\,\varphi^{\#}$$

and analogously we define the arity and domain condition of $\sigma\varepsilon_\Omega^{\#}$ for $\sigma \in \Omega(T_2)$ if there is no $\sigma'' \in \Omega(T_0)$ with $\sigma\varepsilon_\Omega^{\#} = \sigma''\varphi_\Omega\varepsilon_\Omega^{\#}$. Straightforward considerations show that the arity function and the domain conditions are defined independently of the choice of representatives.

Since $\theta(T_0)$, $\theta(T_1)$, $\theta(T_2)$ are hierarchical ep-signatures $\theta(T)$ becomes hierarchical, too.

As next we define the set of axioms $\mathfrak{A}(T)$ by

$$\mathfrak{A}(T) = \mathfrak{A}(T_0)\,\psi \cup \mathfrak{A}(T_1)\,\varphi^{\#} \cup \mathfrak{A}(T_2)\,\varepsilon^{\#}$$

and so $\varepsilon^{\#}: T_2 \to T$, $\varphi^{\#}: T_1 \to T$ evidently become theory-morphisms. Finally we define the set Δ of initial restrictions in T by

$$\Delta = \Delta_1\varphi^{\#} \cup \Delta_2\varepsilon^{\#}$$

with

$$\Delta_1\varphi^{\#} = \{(T', \pi \circ \varphi^{\#}: T'' \to T) \mid (T', \pi: T'' \to T_1) \in \Delta_1\},$$

$$\Delta_2\varepsilon^{\#} = \{(T', \pi \circ \varepsilon^{\#}: T'' \to T) \mid (T', \pi: T'' \to T_2) \in \Delta_2\}.$$

It remains to prove the universal property for the canon morphisms

$$\varphi^{\#}: (T_1, \Delta_1) \to (T, \Delta), \qquad \varepsilon^{\#}: (T_2, \Delta_2) \to (T, \Delta).$$

For this purpose let

$$\varphi^{\#\#}: (T_1, \Delta_1) \to (T^{\#}, \Delta^{\#}), \qquad \varepsilon^{\#\#}: (T_2, \Delta_2) \to (T^{\#}, \Delta^{\#})$$

be canon morphisms with $\varepsilon \circ \varphi^{\#\#} = \varphi \circ \varepsilon^{\#\#}$. Then we obtain functions

$$\delta_S: S(T) \to S(T^{\#}), \qquad \delta_\Omega: \Omega(T) \to \Omega(T^{\#})$$

uniquely determined by

$$\varepsilon_S^{\#} \circ \delta_S = \varepsilon_S^{\#\#}, \qquad \varphi_S^{\#} \circ \delta_S = \varphi_S^{\#\#},$$
$$\varepsilon_\Omega^{\#} \circ \delta_\Omega = \varepsilon_\Omega^{\#\#}, \qquad \varphi_\Omega^{\#} \circ \delta_\Omega = \varphi_\Omega^{\#\#}$$

since

$$\varphi_S^{\#}: S(T_1) \to S(T), \qquad \varepsilon_S^{\#}: S(T_2) \to S(T),$$
$$\varphi_\Omega^{\#}: \Omega(T_1) \to \Omega(T), \qquad \varepsilon_\Omega^{\#}: \Omega(T_2) \to \Omega(T)$$

satisfy the corresponding universal property with respect to

$$\varepsilon_S: S(T_0) \to S(T_1), \qquad \varphi_S: S(T_0) \to S(T_2),$$
$$\varepsilon_\Omega: \Omega(T_0) \to \Omega(T_1), \qquad \varphi_\Omega: \Omega(T_0) \to \Omega(T_2)$$

respectively. One can see easily that $(\delta_S, \delta_\Omega)$ forms a canon morphism

$$\delta: (T, \Delta) \to (T^{\#}, \Delta^{\#}). \qquad \blacksquare$$

The use of the pushout construction for canon morphisms is based on the following theorem proved by in Hupbach (1980) and (1981).

Theorem 4.3.4. *Let the diagram*

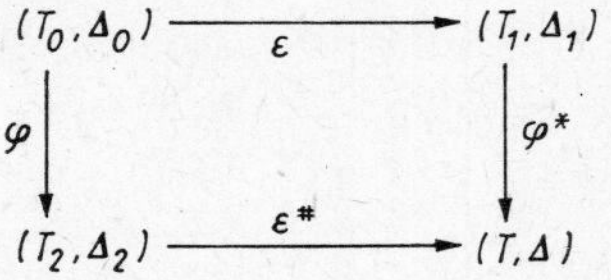

be a pushout of canon morphisms.

(1) *An algebra* $A \in \mathrm{ALG}(T)$ *is a* (T, Δ)*-algebra if and only if* $A\varepsilon^{\#} \in \mathrm{ALG}(T_2, \Delta_2)$ *and* $A\varphi^{\#} \in \mathrm{ALG}(T_1, \Delta_1)$.

(2) *For all algebras* $B \in \mathrm{ALG}(T_1, \Delta_1)$, $C \in \mathrm{ALG}(T_2, \Delta_2)$ *with* $B\varepsilon = C\varphi$ *there is exactly one* (T, Δ)*-algebra, denoted by* $B \oplus C$ *with* $(B \oplus C)\,\varphi^{\#} = B$ *and* $(B \oplus C)\,\varepsilon^{\#} = C$. *For every pair of homomorphisms*

$$f: B_1 \to B_2, \qquad g: C_1 \to C_2$$

with $B_1, B_2 \in \mathrm{ALG}(T_1, \Delta_1)$, $C_1, C_2 \in \mathrm{ALG}(T_2, \Delta_2)$ *and* $f\varepsilon = g\varphi$ *there is exactly one homomorphism*

$$f \oplus g: B_1 \oplus C_1 \to B_2 \oplus C_2$$

with

$$(f \oplus g)\,\varphi^{\#} = f \quad \text{and} \quad (f \oplus g)\,\varphi^{\#} = g.$$

(3) *If* $\varepsilon_0: (T_0, \Delta_0) \to (T_1, \Delta_1)$ *is the inclusion of a subcanon then one can assume without loss of generality that* $\varepsilon^{\#}: (T_2, \Delta_2) \to (T, \Delta)$ *is the inclusion of a subcanon, too.*

(4) *If* $\varepsilon: (T_0, \Delta_0) \to (T_1, \Delta_1)$ *is persistent, then* $\varepsilon^{\#}: (T_2, \Delta_2) \to (T, \Delta)$ *is persistent.*

Proof: (1): Let $A \in \mathrm{ALG}(T)$ and $(T', \pi \circ \varphi^\#: T'' \to T) \in \Delta$ with $(T', \pi: T'' \to T) \in \Delta_1$. Then $A \models (T', \pi \circ \varphi^\#: T'' \to T)$ holds iff $A(\pi \circ \varphi^\#) = (A\varphi^\#)\,\pi$ is freely generated by $(A\varphi^\#)\,\pi\!\downarrow\! T'$, i.e. iff $A\varphi^\# \models (T', \pi: T'' \to T)$. Analogously, $A \models (T', \pi \circ \varepsilon^\#: T'' \to T)$ iff $A\varepsilon^\# \models (T', \pi: T'' \to T)$ for every $(T', \pi: T'' \to T) \in \Delta_2$.

(2): For $B \in \mathrm{ALG}(T_1, \Delta_1)$, $C \in \mathrm{ALG}(T_2, \Delta_2)$ with $B\varepsilon = C\varphi$ we define $B \oplus C$ as follows: $(B \oplus C)_s = B_{s'}$ for $s \in S(T)$ if there is an $s' \in S(T_1)$ with $s = s'\varphi_S^\#$, and $(B \oplus C)_s = C_{s''}$ for $s \in S(T)$ if there is an $s'' \in S(T_2)$ with $s = s''\varphi_S^\#$. Analogously we set $\sigma^{(B\oplus C)} = \sigma_1^B$ if there is a $\sigma_1 \in \Omega(T_1)$ with $\sigma = \sigma_1\varphi_\Omega^\#$ and $\sigma^{(B\oplus C)} = \sigma_2^C$ if there is a $\sigma_2 \in \Omega(T_2)$ with $\sigma = \sigma_2\varepsilon_\Omega^\#$.

In the same way the homomorphism $f \oplus g: B_1 \oplus C_1 \to B_2 \oplus C_2$ can be constructed, i.e. $(f \oplus g)_s = f_{s'}$ if there is an $s' \in S(T_1)$ with $s = s'\varphi_S^\#$, and $(f \oplus g)_s = g_{s''}$ if there is an $s'' \in S(T_2)$ with $s = s''\varepsilon_S^\#$.

(3): We have to prove that $\varepsilon_S^\#: S(T_2) \to S(T)$ and $\varepsilon_\Omega^\#: \Omega(T_2) \to \Omega(T)$ are injective. According to the construction of the sets $S(T)$, $\Omega(T)$ and the definitions of the functions $\varepsilon_S^\#$, $\varepsilon_\Omega^\#$ for $s', s'' \in S(T_2)$

$$s'\varepsilon_S^\# = s''\varepsilon_S^\#$$

holds if there are $s_1, s_2, \ldots, s_n \in S(T_0)$ with

$$s' = s_1\varphi_S, \quad s_1\varepsilon_S = s_2\varepsilon_S,$$

$$s_2\varphi_S = s_3\varphi_S, \quad s_3\varepsilon_S = s_4\varepsilon_S, \ldots, s'' = s_n\varphi_S.$$

Since $\varepsilon_S: S(T_0) \to S(T_1)$ is injective, we obtain $s_1 = s_2$, $s_3 = s_4, \ldots$ and so

$$s' = s_1\varphi_S = s_2\varphi_S = s_3\varphi_S = \cdots = s_n\varphi_S = s''.$$

In the same way the injectivity of $\varepsilon_\Omega^\#: \Omega(T_2) \to \Omega(T)$ follows.

(4): As a preparation we prove the following statement: $A \in \mathrm{ALG}(T, \Delta)$ is freely generated by $A\varepsilon^\#$ according $\varepsilon^\#: T_2 \to T$ if $A\varphi^\#$ is freely generated by $(A\varphi^\#)\varepsilon$ according $\varepsilon: T_0 \to T_1$. For the proof of this assertion let $f: A\varepsilon^\# \to B\varepsilon^\#$ be a homomorphism with $B \in \mathrm{ALG}(T)$. Then there is a homomorphism $f\varphi: (A\varepsilon^\#)\varphi \to (B\varepsilon^\#)\varphi$, and because of $\varphi \circ \varepsilon^\# = \varepsilon \circ \varphi^\#$ it holds that $f\varphi: (A\varphi^\#)\varepsilon \to (B\varphi^\#)\varepsilon$.

Since $A\varphi^\#$ is freely generated by $(A\varphi^\#)\varepsilon$ there is exactly one homomorphism

$$g: A\varphi^\# \to B\varphi^\#.$$

By means of (2) we have a homomorphism

$$f \oplus g: A\varepsilon^\# \oplus A\varphi^\# \to B\varepsilon^\# \oplus B\varphi^\#$$

and because

$$A = A\varepsilon^\# \oplus A\varphi^\# \quad \text{and} \quad B = B\varepsilon^\# \oplus B\varphi^\#$$

there is exactly one homomorphism

$$f \oplus g: A \to B$$

with $(f \oplus g)\, \varepsilon^{\#} = f$. This proves that A is freely generated by $A\varepsilon^{\#}$ according to $\varepsilon^{\#}: T_2 \to T$.

Next we prove condition (1) of Definition 4.3.1. Let $B \in \mathrm{ALG}(T_2, \Delta_2)$ be given. Then $B\varphi \in \mathrm{ALG}(T_0, \Delta_0)$ and the persistence of $\varepsilon: (T_0, \Delta_0) \to (T_1, \Delta_1)$ implies the existence of an algebra $C \in \mathrm{ALG}(T_1, \Delta_1)$ such that $C\varepsilon = B\varphi$ and C is freely generated by $C\varepsilon$.

Because of $B \in \mathrm{ALG}(T_2, \Delta_2)$, $C \in \mathrm{ALG}(T_1, \Delta_1)$ and $B\varphi = C\varepsilon$,

$$B \oplus C \in \mathrm{ALG}(T, \Delta)$$

holds according (2).

Since $(B \oplus C)\varepsilon^{\#} = B$ and since $(B \oplus C)\, \varphi^{\#} = C$ is freely generated by $C\varepsilon = B\varphi$ the statement proved as preparation yields that $B \oplus C$ is freely generated by B and so condition (1) is satisfied. For the proof of condition (2) of Definition 4.3.1 we assume $A \in \mathrm{ALG}(T, \Delta)$. Then $A\varphi^{\#} \in \mathrm{ALG}(T_1, \Delta_1)$ and since $\varepsilon: (T_0, \Delta_0) \to (T_1, \Delta_1)$ is persistent, $A\varphi^{\#}$ is freely generated by $(A\varphi^{\#})\varepsilon = (A\varepsilon^{\#})\varphi$. A further application of the statement proved above implies that A is freely generated by $A\varepsilon^{\#}$ according to $\varepsilon^{\#}: T_2 \to T$ and so condition (2) is also satisfied. ■

The following theorem proves that parameter instantiation commutes with recursiveness.

Theorem 4.3.5. *Let the diagram*

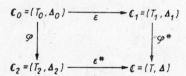

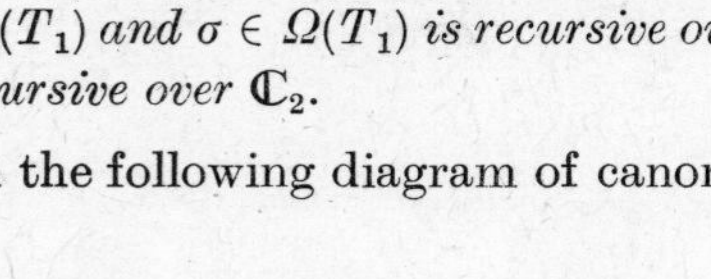

be a pushout of canon morphisms, where $\varepsilon: \mathbb{C}_0 \to \mathbb{C}_1$ *is an inclusion of a subcanon (and hence we can assume that* $\varepsilon^{\#}: \mathbb{C}_2 \to \mathbb{C}$ *is an inclusion of a subcanon, too). If* $s \in S(T_1)$ *and* $\sigma \in \Omega(T_1)$ *is recursive over* $\mathbb{C}_0$*, then* $s\varphi^{\#} \in S(T)$ *and* $\sigma\varphi^{\#} \in \Omega(T)$ *is recursive over* $\mathbb{C}_2$.

Proof. We start with the following diagram of canon morphisms:

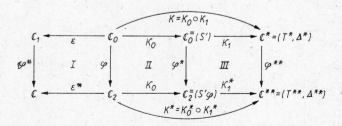

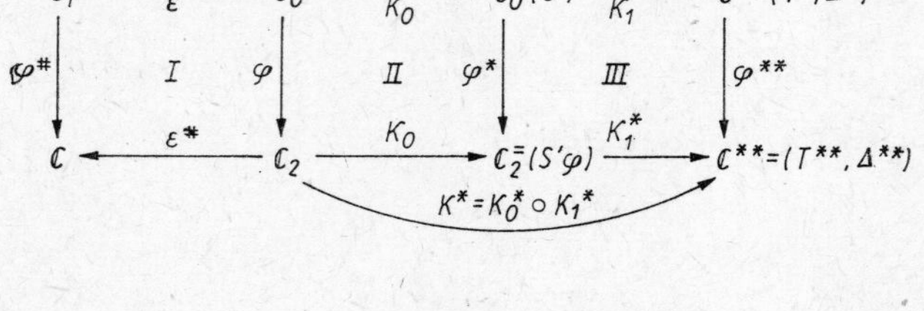

In this diagram the squares I, II, and III are pushouts, where I is the given one, II results from the enrichment of $\mathbb{C}_0$ by S'-identies, and III results from the persistent enlargement $\varkappa_1: \mathbb{C}_0^{=}(S') \to \mathbb{C}^*$. S' and $\mathbb{C}^*$ may be given corresponding to the recursiveness of $s \in S(T_1)$ over $\mathbb{C}_0$. This implies further the existence of $s^* \in S(T^*)$ and of an operator $s^*\text{-eq}: s^*s^* \to \text{Bool}$ satisfying the conditions of Definition 4.3.2.

Due to Theorem 4.3.4, (4) the pushout III gives a persistent enlargement $\varkappa_1^*: \mathbb{C}_2^{=}(S'\varphi) \to \mathbb{C}^{**}$. In the following we will show that $S'\varphi \subseteq S(T_2)$, $\varkappa_1^*: \mathbb{C}_2^{=}(S'\varphi) \to \mathbb{C}^{**}$, $s^*\varphi^{**} \in S(T^{**})$ and $(s^*\text{-eq})\,\varphi^{**} \in \Omega(T^{**})$ satisfy all conditions necessary for the recursiveness of $s\varphi^\# \in S(T)$ over $\mathbb{C}_2 \subseteq \mathbb{C}$.

Let $A \in \text{ALG}(\mathbb{C})$. Then we have $A\varphi^\# \in \text{ALG}(\mathbb{C}_1)$ and $A\varepsilon^\# \in \text{ALG}(\mathbb{C}_2)$ with $(A\varphi^\#)\varepsilon = (A\varepsilon^\#)\varphi$. Since s is recursive over $\mathbb{C}_0$ there is an algebra $B \in \text{ALG}(\mathbb{C}^*)$ so that $(A\varphi^\#)\varepsilon = B\varkappa$, $B_{s^*} = (A\varphi^\#)_s = A_{s\varphi^\#}$, and $(s^*\text{-eq})^B$: $(A\varphi^\#)_s \times (A\varphi^\#)_s \to \{\text{true,false}\}$ is the identity relation.

For the proof of the recursiveness of $s\varphi^\#$ over $\mathbb{C}_2$ we have to construct a suitable $\bar{B} \in \text{ALG}(\mathbb{C}^{**})$. Since II and III are pushouts the outer diagram II + III is a pushout, too. Now we have $A\varepsilon^\# \in \text{ALG}(\mathbb{C}_2)$ and $B \in \text{ALG}(\mathbb{C}^*)$ satisfying $(A\varepsilon^\#)\,\varphi = B\varkappa$, so that by (2) of Theorem 4.3.4 there is exactly one $\mathbb{C}^{**}$-algebra $(A\varepsilon^\#) \oplus B$, which is the one we are looking for, because of $\big((A\varepsilon^\#) \oplus B\big)\,\varkappa^* = A\varepsilon^\#$,

$$\big((A\varepsilon^\#) \oplus B\big)_{s^*\varphi^{**}} = \big(\big((A\varepsilon^\#) \oplus B\big)\,\varphi^{**}\big)_{s^*} = B_{s^*} = A_{s\varphi^\#},$$

and

$$\begin{aligned}\big((s^*\text{-eq})\,\varphi^{**}\big)^{((A\varepsilon^\#)\oplus B)} &= (s^*\text{-eq})^{((A\varepsilon^\#)\oplus B)\varphi^{**}} \\ &= (s^*\text{-eq})^B: (A_{s\varphi^\#}) \times (A_{s\varphi^\#}) \to \{\text{true,false}\}\end{aligned}$$

is the identity relation.

In just the same way one proves the recursiveness of $\sigma\varphi^\#$ over $\mathbb{C}_2$. ■

5. Canons of behaviour

As mentioned in Section 1.1, the semantics of an abstract data Type is given by a class of behaviourally equivalent algebras. Till now behavioural equivalence was formally represented by isomorphisms of algebras. However, in several papers it was explained that the formal representation of behavioural equivalence by isomorphisms of algebras does not fit with the intuitive understanding of data abstraction, see GIARRANTANA et al. (1976), WAND (1979), HORNUNG and RAULEFS (1980), BERGSTRA and MAYER (1981). However, the papers of final or terminal semantics are not exactly on this line, since they also use isomorphism classes, now containing the final or terminal algebra. The motivation for the choice of final algebras seems in our opinion to be strongly connected with the ideas behind observability, since final algebras correspond to automata with minimalized sets of internal states.

The inadequacy of isomorphism classes can be demonstrated by the fact that the most convenient realization of stacks, namely the realization by pointed arrays in which a stack is represented as an array together with a pointer to the top element, is isomorphic neither to the initial algebra nor to the final algebra with respect to the usually given set of defining equations.

Let us consider a specification of stacks, without error values, where the emtpy stack can be neither popped nor topped.

STACK **is** DATA **with definition**
  sorts Stack
  oprn empty → Stack
    push(Stack,Data) → Stack
    top(Stack) ⇸ Data
    pop(Stack) ⇸ Stack
  axioms s: Stack, d: Data
    top(push(s,d)) = d
    pop(push(s,d)) = s
end STACK

and

DATA **is requirement**
  sorts Data
end DATA.

What is wrong with the realization of stacks by (array, pointer)-pairs? In this realization the pop-operation moves down the pointer by one, say from $n + 1$ to n, but it does not change the value '$a(n + 1)$' of the array 'a'.

This is not necessary, since a value can only be read by an index equal to the pointer value, and the pointer can only be moved up by means of the push-operation which simultaneously overwrites the previously written value in the array. But, in this case the required equation 'pop(push(s,d)) $= s$' is not satified in general. However, pop(push(s,d)) and s represent behaviourally equivalent internal states of a stack, if we extend the notion of behaviourally equivalent states in automata to stacks. Now one can say that the realization of stacks by pointed arrays satisfies the equation 'pop(push(s,d)) $= s$' not identically, but behaviourally, since the resulting values of the left-hand side and of the right-hand side are not identical but behaviourally equivalent. This observation becomes fundamental for the generalization of the semantics of initial canons to behavioural semantics.

The following specification will demonstrate that we have not only to extend the semantics of specifications, but that we have also to provide for additional syntactical tools. Let us consider sequences of arbitrary elements:

SEQUENCE **is** DATA **with definition**
  sorts Seq
  oprn nil $\to$ Seq
    cons(Seq,Data) $\to$ Seq
    head(Seq) $\circ\!\!\to$ Data
    tail(Seq) $\circ\!\!\to$ Seq
  axioms s: Seq, d: Data
    head(cons(s,d)) $= d$
    tail(cons(s,d)) $= s$
end SEQUENCE

The following renaming

Seq $\mapsto$ Stack, nil $\mapsto$ empty, cons $\mapsto$ push, head $\mapsto$ top, tail $\mapsto$ pop

transforms the specification SEQUENCE into the specification STACK.

Since any kind of semantics of a specification has to be invariant under renaming, this would mean that sequences and stacks are equal polymorphic data types. This statement contradicts the intuitive understanding of these two data types. Intuitively, sequences are considered as data like integers or truth values, whereas stacks are considered as some kind of machines with internal states. In this sense stacks are used to store data by means of the push operation and to give back data by means of the top operation. This contradiction is not only an effect of the chosen sets of operations. Even with these set of operations both data types are semantically quiete distinct. This means that we have to enrich the syntactical tools of specifications by those that allow descriptions of the differences of both these data types.

It turns out that it is sufficient to divide up the set of sort names in visible (or input/output-) sorts and in hidden (or state-) sorts. These are the only additional syntactical tools. On the basis of these two kinds of sort names we extend from the theory of abstract automate the notion of behaviourally equivalent internal states, and we introduce behavioural satisfaction of elementary implications. We can prove the existence of initial algebras in model classes of elementary implications with respect to behavioural satisfaction. This result guarantees the existence of left-adjoint junctors for every forgetful functor between those model classes. These are the fundamental facts that allow the generalization of initial restrictions to initial restrictions of behaviour and of initial canons to canons of behaviour.

The definition of behavioural equivalence of elements of algebras, the definition of behaviourally equivalent algebras, the interrelations of these both notions, and the existence of initial algebras in behaviourally closed model classes are the subjects of Section 5.1.

In Section 5.2 we extend the concepts of initial restrictions to behavioural semantics, we define canons of behaviour, and we introduce canon morphisms between canons of behaviour.

Section 5.3 deals with problems of computability in canons of behaviour.

Chapter 5 can be considered as a proper extension of the preceding chapters, since the theory developed so far can be characterized by the fact that all sort names are visible ones.

As distinct to some other approaches dealing with observability and parameterized data types, see Broy, Wirsing (1982), Ganzinger (1983) and Kamin and Archer (1984), we do not assume that all parameter types behave towards the polymorphic type like visible types. This concept makes it much more difficult to allow that an instantiation of the parameter canon may be a canon which is not totally restricted, i.e., the parameter types are only partially instantiated to ADTs. Since parameter instantiation is the main tool for stepwise refinement, partial instantiation of parameter canons is basic for top down design. Broy and Wirsing (1981) and (1982) also deal with observable or extensional equivalence of initial and weakly terminal partial algebras. The remarks in Section 4.1 concerning the relation of hierarchical data types and canons are valid for behavioural semantics, too.

A very interesting approach to behavioural semantics of parameterized data types was developed by Sannella, Tarlecki, and Wirsing on the basis of an abstraction operation in a kernel language for specifications and implementations of abstract data types, where the semantics of the kernel language itself is parameterized by an institution in the sense of Goguen and Burstall (1984); see Sannella and Wirsing (1983), Sannella and Tarlecki (1984), (1985 b).

The paper whose approach relates most closely to ours is that of Goguen and MESEGUER (1982). We understand Chapter 5 as an extension of that paper from algebraic theories with total operations to algebraic theories with partial operations, and to initially restricting theories with partial operations. They developed thoroughly the strong connections to modularization of complex software, and they developed this approach to the concept of *parameterized programming* (see GOGUEN (1984)) and derived suggestions concerning libraries for Ada program development (see GOGUEN (1984b)). In this approach partiality is treated by error values and by subsorts.

5.1. Behavioural validity of elementary implications

There are several generalizations of the equivalence of internal states of an automata to observable or behavioural equivalence of elements of many-sorted algebras. Some of them have been compared by HUPBACH and REICHEL (1983). It turns out that all considered concepts coincide on finitely generated many-sorted algebras, i.e. on many-sorted algebras without proper subalgebras, and that the concepts split up in general. Since we are concerned with specifications of polymorphic data types, we cannot work with only finitely generated algebras. Observability has so far mainly been considered for total algebras, for finitely generated partial algebras, or only semantically, i.e. only on the level and in terms of algebras. As sketched above we will follow the basic intuition of the institution concept in GOGUEN, BURSTALL (1984) and we will introduce behavioural validity of conditional equations, which requires the concept of behaviourally equivalent elements of an algebra, and we will show that there is a corresponding concept of behavioural equivalence of algebras. This approach requires the proof of the compatibility of the corresponding notions on both levels; see Theorems 5.1.8 and 5.1.13.

Additionally, we hope that the notion of behavioural validity of conditional equations will essentially support the investigation of operational semantics of behavioural specifications, but this topic is beyond the range of this book.

Many of the problems we are concerned with in this section arise from the existence of partial operations. If one has in mind the problems arising in the theory of automata if one extends the universal state minimalization from total automata to partial ones, then one can not expect that these concepts can simply be generalized from total algebras to equationally partial algebras.

Our concept of observability is based on the division of the set of sort names in visible and hidden sort names. In the case of automata the input alphabet and the output alphabet are visible sorts and the set of states is the only hidden sort.

Definition 5.1.1. An *I/O-signature*

$$\Sigma[I] = (S, I, \alpha\colon \Omega \to S^* \times S)$$

is given by a signature $\Sigma = (S, \alpha: \Omega \to S^* \times S)$ and by a distinguished subset $I \subseteq S$ of *input/output* or *visible* sort names. Any sort name $s \in S$ with $s \notin I$ is called a *hidden* sort name. Σ will denote the underlying signature of an I/0-signature $\Sigma[I]$.

Since we are extending concepts of the theory of automata, we call a model of an I/0-signature $\Sigma[I]$

$$M = \big((M_s \mid s \in S), (\sigma^M \mid \sigma \in \Omega)\big)$$

a $\Sigma[I]$-*machine*. Formally, a $\Sigma[I]$-machine M is nothing more than a partial Σ-algebra with distinguished visible carrier sets M_s, $s \in I$.

$\Sigma[I]$-*homomorphisms* $f: M \to M'$ between $\Sigma[I]$-machines are the homomorphisms between the partial Σ-algebras.

$\mathfrak{M}(\Sigma[I])$ denotes the class of all $\Sigma[I]$-machines. ■

Let $\Sigma[I]$ be any I/0-signature and M a $\Sigma[I]$-machine. A term $t \in T(\Sigma, v)_s$ with $s \in I$ can be interpreted as an abstract experiment on $\Sigma[I]$-machines and

$$t^M: M_v \multimap M_s \quad \text{with} \quad v: X \to S$$

then represents an actual experiment on M. Since we are dealing with partial $\Sigma[I]$-machines, not every configuration $\boldsymbol{m} \in M_v$ consisting of input values $x\boldsymbol{m} \in M_{xv}$ if $xv \in I$ and of internal states $x\boldsymbol{m} \in M_{xv}$ if $xv \notin I$, yields a result. If there is no result, we interpret this fact as an infinite running of the experiment t^M started with the configuration $\boldsymbol{m} \in M_v$. Now we can say that internal states $m_1, m_2 \in M_s$, $s \notin I$, are behaviourally equivalent if for every experiment $t^M: M_v \multimap M_{s'}$ and all configuration $\boldsymbol{m}', \boldsymbol{m}'' \in M_v$, which differ only for one $x_0 \in X$ with $x_0 v = s$ and $x_0\boldsymbol{m}' = m_1$, $x_0\boldsymbol{m}'' = m_2$, the $\Sigma[I]$-machine M behaves identical, i.e. $\boldsymbol{m}'t^M$ comes to a result if and only if $\boldsymbol{m}''t^M$ comes to a result, and then both result are equal visible objects.

To reach a precise version of this notion we need some terminology.

Definition 5.1.2. Let be $w: X \to S$ a finite S-sorted system of variables, $m \in \mathfrak{M}(\Sigma[I])$, $m \in \mathfrak{M}_s$ for some $s \in S$, and $x \in X$ with $xw = s$. $t \in T(\Sigma, w)_{s'}$ is called an *output term* if $s' \in I$. Additionally we set:

(i) $w - x: X - \{x\} \to S$ is the finite S-sorted system of variables that results form $w: X \to S$ by removing $x \in X$.

(ii) If $\boldsymbol{m} \in M_{w-x}$ and $m \in M_{xv}$, then

$$\boldsymbol{m} + (x = m) \in M_w$$

denotes the extended assignment, as indicated by the introduced notation.

(iii)
$$t^M_{x=m}: M_{w-x} \multimap M_{s'}$$

denotes the partial induced polynomial function with $\boldsymbol{m} \in \mathrm{dom}(t^M_{x=m})$ if and only if $\boldsymbol{m} + (x = m) \in \mathrm{dom}(t^M)$ and

$$\boldsymbol{m}t^M_{x=m} = \big(\boldsymbol{m} + (x = m)\big)\, t^M. \qquad ■$$

Notice, an output term $t \in T(\Sigma, w)_s$, $s \in I$, is always associated with a finite S-sorted system of variables $w: X \to S$.

Analogously to the notation of operators we will use the notation $t: w \to s$ instead of $t \in T(\Sigma, w)_s$ if Σ is evident from context.

Definition 5.1.3. Let A be a $\Sigma[I]$-machine. The elements $a, b \in A_s$, $s \in S$, are called *behaviourally I-equivalent*, in symbols

$$a \equiv b \bmod I$$

if for every output term $t: w \to s'$, with $w: X \to S$ so that there is an $x_0 \in X$ with $x_0 w = s$, the following conditions are satisfied:

(i) $\mathrm{dom}(t^A_{x_0=a}) = \mathrm{dom}(t^A_{x_0=b})$

(ii) $\boldsymbol{a}(t^A_{x_0=a}) = \boldsymbol{a}(t^A_{x_0=b})$ for all $\boldsymbol{a} \in \mathrm{dom}(t^A_{x_0=a}) \subseteq A_{w-x_0}$.

The elements $a, b \in A_s$, $s \in S$, are said to be *I-distinguishable* if there is an output term $t: w \to s'$, $w: X \to S$, if there is an $x_0 \in X$ with $x_0 w = s$, and if there is an assignment

$$\boldsymbol{a} \in \mathrm{dom}(t^A_{x_0=a}) \cap \mathrm{dom}(t^A_{x_0=b})$$

with

$$\boldsymbol{a}(t^A_{x_0=a}) \neq \boldsymbol{a}(t^A_{x_0=b}).$$ ■

As distinct from the situation of total $\Sigma[I]$-machines, as considered by GIARRATANA et al. (1976), the concepts of I-equivalence and I-distinguishability are no longer complementary concepts for partial $\Sigma[I]$-machines.

Even if the following corollary is a simple consequence of the preceding definition, it states an important property of I-equivalence.

Corollary 5.1.4. *For every I/O-signature $\Sigma[I]$ and every $\Sigma[I]$-machine A elements of visible sorts are behaviourally I-equivalent if and only if they are equal.*

This is caused by the fact that for a visible sort $s \in I$ any variable $x \in X$ of a finite S-sorted system of variables $w: X \to S$ with $xw = s$ itself is an output term which is naturally able to distinguish any two different elements $a, b \in A_{xw}$.

Due to this corollary the theory developed in the preceding chapters is preserved as the special case in which all sort names are visible ones.

It is easy to prove that I-equivalence of elements is really an equivalence relation in each carrier set A_s, $s \in S$, for every $\Sigma[I]$-machine A.

Next we generalize the concept of behaviourally equivalent automata. The behaviourally equivalence of $\Sigma[I]$-machines is based on the following definition:

Definition 5.1.5. A $\Sigma[I]$-homomorphism $f: A \to B$ is called an *I-reduction* if

(1) $f_s: A_s \to B_s$ is surjective for every sort name $s \in S$;

(2) $f_s: A_s \to B_s$ is additionally injective for every visible sort name $s \in I$;

(3) for all operators $\sigma: v \to s$ of Σ and every $\boldsymbol{a} \in A_v$ the relation $\boldsymbol{a}f_v \in \mathrm{dom}\, \sigma^B$ implies $\boldsymbol{a} \in \mathrm{dom}\, \sigma^A$.

$\Sigma[I]$-machines A, B will be called *I-equivalent*, in symbols

$$A \equiv B \bmod I,$$

if there is a $\Sigma[I]$-machine F and there reductions $r: F \to A$, $r': F \to B$. ■

The definition of equivalence of automata says that automata A, B are equivalent if they can be reduced by identification of equivalent internal states to a common automaton C, i.e. if there are reductions $A \to C$, $B \to C$. From this definition one can prove that automata A, B are equivalent if there is an automaton D that can be reduced to A and B, i.e. if there are reductions $D \to A$, $D \to B$. We start from the second concept to define the equivalence of $\Sigma[I]$-machines since it is easier to prove that pullback constructions preserve reductions (see 5.1.6 below) than to prove the corresponding preservation for pushouts. This result is necessary for the transitivity of behavioural equivalence of $\Sigma[I]$-machines.

Our notion of I-reduction and I-equivalence is not based on the existence of initial states in machines. I-equivalent $\Sigma[I]$-machines A, B should behave equally independent on actual states of A and B in which we start an experiment. We guess that during a top-down design in a very late refinement step initial states can be distinguished. Thus, at the beginning we need behavioural equivalence that can be defined without the use of initial states. The correspondence of this intuition to the I-equivalence of elements of $\Sigma[I]$-machines is more recognizable than to the preceding definition. Theorems 5.1.8 and 5.1.13 however show that also this definition corresponds to our intuition.

Certainly, one can start from different intuitions on behavioural equivalence of many-sorted algebras. In Hupbach, Reichel (1983) it was proved for total algebras that the preceding concept of behavioural equivalence induces the smallest classes of behaviourally equivalent algebras. This fact together with the preceding consideration encourage us to investigate more deeply this kind of observability.

This definition of behavioural equivalence of $\Sigma[I]$-machines is a bit more general than equivalence of automata, since we allow isomorphism of input alphabets and output alphabets respectively, instaed of their identity.

If it was obvious that I-equivalence of elements of $\Sigma[I]$-machines is an equivalence relation on every carrier set of a $\Sigma[I]$-machine, for the I-equivalence of $\Sigma[I]$-machines only symmetry and reflexivity are immediate consequences of the definition. The transitivity becomes a consequence of the next corollary.

Corollary 5.1.6. *If the following diagram*

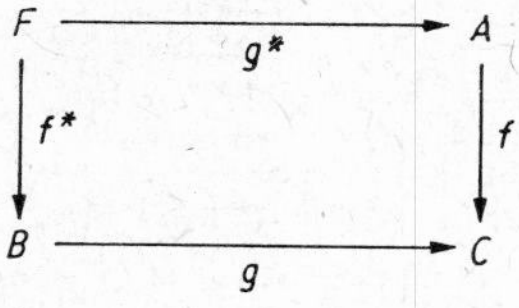

is a pullback diagram of $\Sigma[I]$-homomorphisms, and if $f: A \to C$, $g: B \to C$ are reductions, then the $\Sigma[I]$-homomorphisms $f^*: F \to B$, $g^*: F \to A$ are reductions, too.

Remember, due to elementary category theory (see MacLane (1971)) F is up to isomorphism that subalgebra of the Cartesian product $B \times A$, with $(x, y) \in F$ if and only if $xg = yf$. f^* and g^* are the reductions of the first and of the second projection of $B \times A$, respectively.

According to that construction of the $\Sigma[I]$-machine F and of the $\Sigma[I]$-homomorphisms $f^*: F \to B$, $g^*: F \to A$ the proof of Corollary 5.1.6 is straightforward.

If we assume that A, B and B, C are I-equivalent, then there are $\Sigma[I]$-machines F_1, F_2 and I-reductions $r_1: F_1 \to A$, $q_1: F_1 \to B$, $r_2: F_2 \to B$, $q_2: F_2 \to C$. Following Corollary 5.1.6 we obtain by a pullback a $\Sigma[I]$-machine F and I-reductions $q_1^*: F \to F_2$, $r_2^*: F \to F_1$. Since the composition of I-reductions yields I-reductions $r_2^* \cdot r_1: F \to A$, $q_1^* \cdot q_2: F \to C$ it holds $A \equiv C \bmod I$.

It is obvious that reductions become isomorphisms if all sort names are visible, and that I-equivalence of $\Sigma[I]$-machines coincides with isomorphism in that case.

The next definition introduces behavioural validity of elementary implications in $\Sigma[I]$-machines.

Definition 5.1.7. Let $\Sigma[I]$ be an I/0-signature, M a $\Sigma[I]$-machine, and $v: X \to S$ a (not necessarily finite) S-sorted system of variables.

(1) For terms $t': v \to s$, $t'': v \to s$ we say that

$$M_{(I,v:t' \stackrel{e}{=} t'')} = \{\boldsymbol{m} \in M_v \mid \boldsymbol{m} \in \mathrm{dom}(t'^M) \cap \mathrm{dom}(t''^M),$$
$$\boldsymbol{m}(t'^M) \equiv \boldsymbol{m}(t''^M) \bmod I\}$$

is the set of *solutions of the I-equivalence* $(I, v: t' \stackrel{e}{=} t'')$ on $v: X \to S$. For a set of I-equivalences $(I, v: \{t_j \stackrel{e}{=} r_j \mid j \in J\})$ with $t_j, r_j: v \to s_j$, $s_j \in S$, $j \in J$, we define

$$M_{(I,v:\{t_j \stackrel{e}{=} r_j \mid j \in J\})} = \bigcap_{j \in J} M_{(I,v:t_j \stackrel{e}{=} r_j)} \quad \text{if } J \neq \emptyset,$$

and

$$M_{(I,v:\emptyset)} = M_v \quad \text{if} \quad J = \emptyset.$$

(2) $M \in \mathfrak{M}(\Sigma[I])$ *behaviourally satisfies* the elementary implication

$$(v\colon t_1 \stackrel{e}{=} r_1, \ldots, t_n \stackrel{e}{=} r_n \to t_0 \stackrel{e}{=} r_0),$$

in symbols

$$M \models \operatorname{mod} I\ (v\colon t_1 \stackrel{e}{=} r_1, \ldots, t_n \stackrel{e}{=} r_n \to t_0 \stackrel{e}{=} r_0),$$

if

$$M_{(I,v:t_1 \stackrel{e}{=} r_1, \ldots, t_n \stackrel{e}{=} r_n)} \subseteq M_{(I,v:t_0 \stackrel{e}{=} r_0)}.$$

If the single I-equivalences of a set of I-equivalences of an elementary implication are not of interest we simply write

$$(I, v\colon G), \qquad (v\colon G \to t \stackrel{e}{=} r)$$

respectively. ■

Due to Corollary 5.1.4 this definition generalizes Definition 2.5.1. This generalization can formally be described by

$$M \models \operatorname{mod} S(v\colon G \to t \stackrel{e}{=} r) \quad \text{if and only if} \quad M \models (v\colon G \to t \stackrel{e}{=} r).$$

The following theorem states the compatibility of the I-equivalence of elements of $\Sigma[I]$-machines and of the I-equivalence of $\Sigma[I]$-machines.

Theorem 5.1.8. *If $f\colon A \to B$ is a reduction of $\Sigma[I]$-machines and $(I, v\colon G)$ with $v\colon X \to S$ is a set of I-equivalences, then*

$$\boldsymbol{a} \in A_{(I,v:G)} \quad \textit{if and only if} \quad \boldsymbol{a} f_v \in B_{(I,v:G)}.$$

Proof. If $(I, v\colon G) = (I, v\colon \emptyset)$ then evidently $\boldsymbol{a} \in A_v$ if and only if $\boldsymbol{a} f_v \in B_v$. Now let $(I, v\colon G) \neq (I, v\colon \emptyset)$.

If $\boldsymbol{a} \in A_{(I,v:G)}$ then, for every $(I, v\colon r_1 \stackrel{e}{=} r_2) \in (I, v\colon G)$,

$$\boldsymbol{a} \in \operatorname{dom}(r_1^A) \cap \operatorname{dom}(r_2^A) \quad \text{and} \quad \boldsymbol{a}(r_1^A) \equiv \boldsymbol{a}(r_2^A) \operatorname{mod} I.$$

By induction one can easily show that for every term $t\colon v \to s$ and every homomorphism $f\colon A \to B$ the membership $\boldsymbol{a} \in \operatorname{dom}(t^A)$ implies $\boldsymbol{a} f_v \in \operatorname{dom}(t^B)$. So we obtain $\boldsymbol{a} f_v \in \operatorname{dom}(r_1^B) \cap \operatorname{dom}(r_2^B)$ and it remains to show $\boldsymbol{a} f_v r_1^B \equiv \boldsymbol{a} f_v r_2^B \operatorname{mod} I$. For that reason let any output term $t\colon w \to s'$ be given with $w\colon Y \to S$, Y finite, $s' \in I$ and $y_0 w = s$ for some $y_0 \in Y$, and $r_1, r_2\colon v \to s$.

By definition $\boldsymbol{a}(r_1^A) \equiv \boldsymbol{a}(r_2^A) \operatorname{mod} I$ implies $t^A_{y_0 = \boldsymbol{a}(r_1^A)} = t^A_{y_0 = \boldsymbol{a}(r_2^A)}$. If $b \in B_{w - y_0}$ then there is an $\boldsymbol{a}' \in A_{w - y_0}$ with $\boldsymbol{b} = \boldsymbol{a}'(f_{w - y_0})$.

Due to condition (3) of Definition 5.1.5 and a simple induction, every reduction not only preserves definedness but also reflects definedness. Therefore we obtain $\boldsymbol{a}' \in \operatorname{dom}\left(t^A_{y_0 = \boldsymbol{a}(r_1^A)}\right)$ if and only if $\boldsymbol{a}' + \left(y_0 = \boldsymbol{a}(r_1^A)\right) \in \operatorname{dom}(t^A)$ if and only if

$$\begin{aligned}\left(\boldsymbol{a}' + \left(y_0 = \boldsymbol{a}(r_1^A)\right)\right) f_w &= \boldsymbol{b} + \left(y_0 = \left(\boldsymbol{a}(r_1^A)\right) f_s\right) \\ &= \boldsymbol{b} + \left(y_0 = (\boldsymbol{a} f_v)\,(r_1^B)\right) \in \operatorname{dom}(t^B)\end{aligned}$$

if and only if $\boldsymbol{b} \in \operatorname{dom}\left(t^B_{y_0 = (\boldsymbol{a} f_v)(r_1^B)}\right)$.

Together with

$$\operatorname{dom}\left(t^A_{y_0=\boldsymbol{a}(r_1^A)}\right) = \operatorname{dom}\left(t^A_{y_0=\boldsymbol{a}(r_2^A)}\right)$$

this implies

$$\operatorname{dom}\left(t^B_{y_0=(\boldsymbol{a}f_v)(r_1^B)}\right) = \operatorname{dom}\left(t^B_{y_0=(\boldsymbol{a}f_v)(r_2^B)}\right).$$

Additionally, for every $\boldsymbol{b} \in \operatorname{dom}\left(t^B_{y_0=(\boldsymbol{a}f_v)(r_1^B)}\right)$ it follows that

$$\begin{aligned}\boldsymbol{b}\left(t^B_{y_0=(\boldsymbol{a}f_v)(r_1^B)}\right) &= \left(\boldsymbol{b} + \left(y_0 = (\boldsymbol{a}f_v)\,(r_1^B)\right)\right) t^B = \left(\boldsymbol{a}' + \left(y_0 = \boldsymbol{a}(r_1^A)\right)\right) f_w t^B \\ &= \left(\boldsymbol{a}' + \left(y_0 = \boldsymbol{a}(r_1^A)\right)\right) t^A f_{s'} = \left(\boldsymbol{a}' + \left(y_0 = \boldsymbol{a}(r_2^A)\right)\right) t^A f_{s'} \\ &= \boldsymbol{b}\left(t^B_{y_0=(\boldsymbol{a}f_v)(r_2^B)}\right),\end{aligned}$$

so that $(\boldsymbol{a}f_v)\,(r_1^B) \equiv (\boldsymbol{a}f_v)\,(r_2^B) \bmod I$. Hence, $\boldsymbol{a} \in A_{(I,v:G)}$ implies $\boldsymbol{a}f_v \in B_{(I,v:G)}$.

For the second half of the proof let $\boldsymbol{a}f_v \in B_{(I,v:G)}$ and $(I, v\colon r_1 \stackrel{e}{=} r_2) \in (I, v\colon G)$ an arbitrary element. We will prove $\boldsymbol{a} \in A_{(I,v:G)}$. $\boldsymbol{a}f_v \in B_{(I,v:G)}$ implies $\boldsymbol{a}f_v \in \operatorname{dom}(r_1^B) \cap \operatorname{dom}(r_2^B)$ and $(\boldsymbol{a}f_v)\,(r_1^B) \equiv (\boldsymbol{a}f_v)\,(r_2^B) \bmod I$.

Since $f\colon A \to B$ reflects definedness, we have $\boldsymbol{a} \in \operatorname{dom}(r_1^A) \cap \operatorname{dom}(r_2^A)$, and the relation $\boldsymbol{a}(r_1^A) \equiv \boldsymbol{a}(r_2^A) \bmod I$ remains to be proved.

Let $r_1, r_2\colon v \to s$, $t\colon w \to s'$ be any output term with $w\colon Y \to S$, Y finite, $y_0 \in Y$ with $y_0 w = s$, and $s' \in I$.

In the same way as above one proves $\operatorname{dom}\left(t^A_{y_0=\boldsymbol{a}(r_1^A)}\right) = \operatorname{dom}\left(t^A_{y_0=\boldsymbol{a}(r_2^A)}\right)$ since $\operatorname{dom}\left(t^B_{y_0=(\boldsymbol{a}f_v)(r_1^B)}\right) = \operatorname{dom}\left(t^B_{y_0=(\boldsymbol{a}f_v)(r_2^B)}\right)$.

If $\boldsymbol{a}' \in \operatorname{dom}\left(t^A_{y_0=\boldsymbol{a}(r_1^A)}\right)$ then

$$\begin{aligned}\left(\boldsymbol{a}'\left(t^A_{y_0=\boldsymbol{a}(r_1^A)}\right)\right) f_{s'} &= \left(\boldsymbol{a}' + \left(y_0 = \boldsymbol{a}(r_1^A)\right)\right) t^A f_{s'} = \left(\boldsymbol{a}' + \left(y_0 = \boldsymbol{a}(r_1^A)\right)\right) f_w t^B \\ &= \left(\boldsymbol{a}' f_{w-y_0} + \left(y_0 = \left(\boldsymbol{a}(r_1^A)\right) f_s\right)\right) t^B \\ &= \left(\boldsymbol{a}' f_{w-y_0} + \left(y_0 = (\boldsymbol{a}f_v)\,(r_1^A)\right)\right) t^B \\ &= (\boldsymbol{a}' f_{w-y_0})\, t^B_{y_0=(\boldsymbol{a}f_v)(r_1^B)} \stackrel{(*)}{=} (\boldsymbol{a}' f_{w-y_0})\, t^B_{y_0=(\boldsymbol{a}f_v)(r_2^B)} \\ &= \left(\boldsymbol{a}'\left(t^A_{y_0=\boldsymbol{a}(r_2^A)}\right)\right) f_{s'}\end{aligned}$$

where equation $(*)$ holds because of $\boldsymbol{a}f_v \in B_{(I,v:r\stackrel{e}{=}r_2)}$.

Because of $s' \in I$ and condition (2) of Definition 5.1.5 the mapping $f_{s'}\colon A_{s'} \to B_{s'}$ is injective so that

$$\boldsymbol{a}'\left(t^A_{y_0=\boldsymbol{a}(r_1^A)}\right) = \boldsymbol{a}'\left({}^A_{y_0=\boldsymbol{a}(r_2^A)}\right).$$

This proves $\boldsymbol{a}(r_1^A) \equiv \boldsymbol{a}(r_2^A) \bmod I$ and so $\boldsymbol{a} \in A_{(I,v:G)}$. ■

This theorem implies that reductions preserve and reflect I-equivalence of elements, i.e., if $r\colon A \to B$ is a reduction, and $a, b \in A_s$, then $a \equiv b \bmod I$ if and only if $ar_s \equiv br_s \bmod I$.

A further noteworthy consequence of Theorem 5.1.8 is the following corollary.

Corollary 5.1.9. *Let* $r: A \to B$ *be any reduction of* $\Sigma[I]$*-machines and* $(v: t_1 \overset{e}{=} r_1, \ldots, t_n \overset{e}{=} r_n \to t_0 \overset{e}{=} r_0)$ *any elementary implication. Then*

$$A \models \mathrm{mod}\, I(v: t_1 \overset{e}{=} r_1, \ldots, t_n \overset{e}{=} r_n \to t_0 \overset{e}{=} r_0)$$

if and only if

$$B \models \mathrm{mod}\, I(v: t_1 \overset{e}{=} r_1, \ldots, t_n \overset{e}{=} r_n \to t_0 \overset{e}{=} r_0).$$

Proof: If $A \models \mathrm{mod}\, I(v: t_1 \overset{e}{=} r_1, \ldots, t_n \overset{e}{=} r_n \to t_0 \overset{e}{=} r_0)$ then

$$A_{(I,v:\{t_1 \overset{e}{=} r_1, \ldots, t_n \overset{e}{=} r_n\})} \subseteq A_{(I,v:t_0 \overset{e}{=} r_0)}. \qquad (\#)$$

Let $\boldsymbol{b} \in B_{(I,v:\{t_1 \overset{e}{=} r_1, \ldots, t_n \overset{e}{=} r_n\})}$ and $\boldsymbol{a} \in A_v$ with $\boldsymbol{b} = \boldsymbol{a}r_v$. The preceding theorem implies $\boldsymbol{b} \in B_{(I,v:\{t_1 \overset{e}{=} r_1, \ldots, t_n \overset{e}{=} r_n\})}$ if and only if $\boldsymbol{a} \in A_{(I,v:\{t_1 \overset{e}{=} r_1, \ldots, t_n \overset{e}{=} r_n\})}$. Because of $(\#)$ above we obtain $\boldsymbol{a} \in A_{(I,v:t_0 \overset{e}{=} r_0)}$, and a second application of the preceding theorem yields $\boldsymbol{b} = \boldsymbol{a}r_v \in B_{(I,v:t_0 \overset{e}{=} r_0)}$, so that

$$B \models \mathrm{mod}\, I\ (v: t_1 \overset{e}{=} r_1, \ldots, t_n \overset{e}{=} r_n \to t_0 \overset{e}{=} r_0)$$

is proved. The converse direction follows in the same way. ■

Corollary 5.1.9 tells us that it is impossible to describe differences of I-equivalent $\Sigma[I]$-machines in terms of elementary implications with behavioural satisfaction.
If $\mathfrak{A}$ is any set of elementary implications,

$$\mathfrak{M}(\Sigma[I], \mathfrak{A})$$

denotes the class of all $\Sigma[I]$-machines which behaviourally satisfy each elementary implication in $\mathfrak{A}$. Obviously, $\mathfrak{M}(\Sigma[I], \mathfrak{A})$ is closed with respect to I-equivalent $\Sigma[I]$-machines.

In most papers, dealing with behavioural equivalence of algebras, the closure with respect to I-equivalence of the class ALG($\mathfrak{A}$), i.e., of the class of all $\Sigma[I]$-machines satisfying identically each elementary implication in $\mathfrak{A}$, is considered. If we denote this class with $\mathfrak{M}^{\#}(\Sigma[I], \mathfrak{A})$, then

$$\mathfrak{M}^{\#}(\Sigma[I], \mathfrak{A}) \subseteq \mathfrak{M}(\Sigma[I], \mathfrak{A}).$$

Now the question arises, of whether the equality holds, and if so behavioural satisfaction allows finite axiomatizations of those classes $\mathfrak{M}^{\#}(\Sigma[I], \mathfrak{A})$. At the moment we are not able to answer this question. The generalization of $\mathfrak{A}$-algebras $F(\mathfrak{A}, G, v)$ freely generated in ALG($\mathfrak{A}$) by a set $(v: G)$ of existence equations to $\Sigma[I]$-machines

$$F(I, \mathfrak{A}, G, v) \in \mathfrak{M}(\Sigma[I], \mathfrak{A})$$

freely generated in $\mathfrak{M}(\Sigma[I], \mathfrak{A})$ by a set $(I, v: G)$ of I-equivalences will become a very powerful tool which also will allow a positive answer to the preceding question.

Definition 5.1.10. A $\Sigma[I]$-macheine $F \in \mathfrak{M}(\Sigma[I])$ *is freely generated in* $\mathfrak{M}(\Sigma[I], \mathfrak{A})$, where $\mathfrak{A}$ is a set of elementary implications, *by a set* $(I, v\colon G)$ *of I-equivalences,* if

(1) $F \in \mathfrak{M}(\Sigma[I], \mathfrak{A})$,

(2) there is a solution $\boldsymbol{e} \in F_{(I,v:G)}$ so that for every $\Sigma[I]$-machine $A \in \mathfrak{M}(\Sigma[I], \mathfrak{A})$ and every solution $\boldsymbol{a} \in A_{(I,v:G)}$ there is exactly one $\Sigma[I]$-homomorphism

$$f\colon F \to A \quad \text{with} \quad \boldsymbol{e}f_v = \boldsymbol{a}. \qquad \blacksquare$$

The definition scheme is the same as that of Definition 3.2.1. We have only substituted the model class ALG($\mathfrak{A}$) by $\mathfrak{M}(\Sigma[I], \mathfrak{A})$ and substituted the freely generating set $(v\colon G)$ of existence equations by a set $(I, v\colon G)$ of I-equivalences. These substitutions are natural in the case of behavioural semantics, since we will formulate properties of $\Sigma[I]$-machines only in terms of I-equivalences and elementary implications behaviourally satisfied.

It is evident that a $\Sigma[I]$-machine $F \in \mathfrak{M}(\Sigma[I], \mathfrak{A})$ freely generated by $(I, v\colon G)$ is uniquely determined up to isomorphisms if it exists. Therefore we will use the notation

$$F(I, \mathfrak{A}, G, v)$$

for that kind of $\Sigma[I]$-machine, extending the notation $F(\mathfrak{A}, G, v)$ of Section 3.2.1.

As a first step in proving the existence of $\Sigma[I]$-machines $F(I, \mathfrak{A}, G, v)$, we show the soundness of the inference rules of Definition 3.1.2 with respect to behavioural satisfaction.

Theorem 5.1.11. *Let $\mathfrak{A}$ be a set of elementary implications and $(v\colon G \to t = r)$ an elementary implication. If*

$$\mathfrak{A} \vdash (v\colon G \to t \overset{\text{e}}{=} r),$$

i.e. $(v\colon G \to t \overset{\text{e}}{=} r)$ *is $\mathfrak{A}$-derivable in the sense of Definition* 3.1.2, *then for every $\Sigma[I]$-machine $A \in \mathfrak{M}(\Sigma[I], \mathfrak{A})$ implies*

$$A \models \operatorname{mod} I(v\colon G \to t \overset{\text{e}}{=} r).$$

Proof: Since I-equivalence of elements in $\Sigma[I]$-machines gives an equivalence relation in every carrier set, the assertion holds for every elementary implication REF_s, SYM_s, and TRA_s. Since output terms are closed under substitution of variables by terms, a straightforward induction proves that the assertion is true for the elementary implications $\text{SUB}_{\sigma,j}$. The derivation rules DI and DII cause no problems.

For the derivation rule DIII let $A \in \mathfrak{M}(\Sigma[I], \mathfrak{A})$ and $\boldsymbol{a} \in A_{(I,v:G)}$. Condition (b) implies $\boldsymbol{a} \in \operatorname{dom}(y\boldsymbol{t})^A$ for all $y \in Y$, so that we can define $\boldsymbol{b} \in A_w$ by $y\boldsymbol{b} = \boldsymbol{a}(y\boldsymbol{t})^A$ for every $y \in Y$. Due to condition (c) we obtain

$\boldsymbol{b} \in A_{(I,w:H)}$ and therefore $\boldsymbol{b}(t_1^A) \equiv \boldsymbol{b}(t_2^A) \bmod I$ according to condition (a). Because of $\boldsymbol{b}(t_1^A) = \boldsymbol{a}(t_1\tilde{\boldsymbol{t}})^A$ and $\boldsymbol{b}(t_2^A) = \boldsymbol{a}(t_2\tilde{\boldsymbol{t}})^A$, where $t_1\tilde{\boldsymbol{t}}$ and $t_2\tilde{\boldsymbol{t}}$ denotes the result of the substitution of the variables $x \in X$ in t_1 and t_2 according to the assignment $\boldsymbol{t} \in T(\Sigma, v)_w$, we obtain

$$\boldsymbol{a}(t_1\tilde{\boldsymbol{t}})^A \equiv \boldsymbol{a}(t_2\tilde{\boldsymbol{t}})^A \bmod I.$$

This proves the soundness of DIII. ■

The following construction of $\Sigma[I]$-machines $F(I, \mathfrak{A}, G, v)$ can be motivated as follows. Elementary implication with behavioural semantics and also I-equivalences can only cause identifications of visible objects, i.e. of elements in some M_s with $s \in I$. In terms of the theory of automata freely generated means to be a reachable $\Sigma[I]$-machine in $\mathfrak{M}(\Sigma[I], \mathfrak{A})$ of *maximal redundancy*, i.e. internal state represents its complete genesis. Algebraically this idea can be realized representing internal states by terms $t: w \to s$, with $s \notin I$, which contain no proper subterms that produce visible values. Thus, only on the levels of the tree, representing the term t, visible elements may occur. One can imagine these trees as *I-pruned trees*. In the case of stacks, where 'Data' is the only visible sort name, and 'Stack' the only hidden one, the terms

empty,
pop(push(push(empty, d), d')),
push(pop(push(empty, d)), d')

are examples of Data-pruned trees, whereas

push(empty,top(push(empty, d)))

is not Data-pruned. The latter term contains a proper subterm producing a visible element of sort 'Data'. But in $F(I, \mathfrak{A}, G, v)$ the genesis of visible elements is hidden, and so this subterm has to be cut out, or has to be reduced to the value represented by this subterm.

This idea can very well be formally realized within the calculus of Sections 3.1 and 3.2. $F(I, \mathfrak{A}, G, v)$ can be constructed as a partial Σ-algebra freely generated in ALG(Σ) by the set of all $\mathfrak{A}$-consequences of $(v: G)$ affecting visible elements. For that reason we define for every set of I-equivalences $(I, v: G)$ the following set of existence equations:

$\big(v: G(I, \mathfrak{A})\big) = \{(v: t \overset{e}{=} r) \mid t, r: v \to s,\ \ s \in S,\ \ \mathfrak{A} \vdash (v: G \to t \overset{e}{=} r),$ and if $s \notin I$ then additionally t and r have to be identical terms$\}$.

The latter condition in the construction of $\big(v: G(I, \mathfrak{A})\big)$ is caused by partiality. This condition generates those terms which represent internal states. This condition additionally guarantees that $\big(v: G(I, \mathfrak{A})\big)$ identifies terms only if they produce visible elements.

Theorem 5.1.12. *For every I/0-signature $\Sigma[I]$, every set $\mathfrak{A}$ of elementary implications, with premises that produce values in visible sorts, and every set $(I, v\colon G)$ of I-equivalence, there exists a $\Sigma[I]$-machine*

$$F(I, \mathfrak{A}, G, v) \in \mathfrak{M}(\Sigma[I], \mathfrak{A})$$

freely generated by $(I, v\colon G)$

Proof: According to the sketched idea of the proof we show that

$$F(\emptyset, G(I, \mathfrak{A}), v) \in \mathrm{ALG}(\Sigma)$$

is freely generated in $\mathfrak{M}(\Sigma[I], \mathfrak{A})$ by $(I, v\colon G)$.

Because of Theorem 5.1.11 we obtain for every $A \in \mathfrak{M}(\Sigma[I], \mathfrak{A})$ and every $(v\colon t \stackrel{e}{=} r) \in (v\colon G(I, \mathfrak{A}))$, $A \models \bmod I\ (v\colon G \to t \stackrel{e}{=} r)$ and therefore

$$A_{(I,v:G)} \subseteq A_{(I,v:t \stackrel{e}{=} r)}.$$

A consequence of the construction of $(v\colon G(I, \mathfrak{A}))$ and of Corollary 5.1.4 for every $(v\colon t \stackrel{e}{=} r) \in (v\colon G(I, \mathfrak{A}))$ is

$$A_{(I,v:t \stackrel{e}{=} r)} = A_{(v:t \stackrel{e}{=} r)}.$$

This implies $\boldsymbol{a} \in A_{(v:G(I,\mathfrak{A}))}$ for every $\boldsymbol{a} \in A_{(I,v:G)}$.

Since $F(\emptyset, G(I, \mathfrak{A}), v)$ is freely generated by $(v\colon G(I, \mathfrak{A}))$ in the class of all Σ-algebras, there is exactly one homomorphism

$$h\colon F(\emptyset, G(I, \mathfrak{A}), v) \to A$$

with $\boldsymbol{e}h_v = \boldsymbol{a}$, where $\boldsymbol{e} \in F(\emptyset, G(I, \mathfrak{A}), v)_v$ with $x\boldsymbol{e} = [x, G(I, \mathfrak{A})]$ for every $x \in X$.

It remains to show that $F(\emptyset, G(I, \mathfrak{A}), v)$ is in $\mathfrak{M}(\Sigma[I], \mathfrak{A})$. For that reason we consider the Σ-algebra $F(\mathfrak{A}, G, v)$ freely generated in $\mathrm{ALG}(\mathfrak{A})$ by $(v\colon G)$. Because of $F(\mathfrak{A}, G, v) \in \mathfrak{M}(\Sigma[I], \mathfrak{A})$ and $\boldsymbol{k} \in F(\mathfrak{A}, G, v)_{(v:G)} \subseteq F(\mathfrak{A}, G, v)_{(I,v:G)}$, with $x\boldsymbol{k} = [x, G]$ for every $x \in X$, there exists exactly one homomorphism

$$f\colon F(\emptyset, G(I, \mathfrak{A}), v) \to F(\mathfrak{A}, G, v)$$

with $\boldsymbol{e}f_v = \boldsymbol{k}$. By induction one can prove for $t\colon v \to s$,

$$[t, G(I, \mathfrak{A})]\, f_s = [t, G].$$

Theorem 5.1.12 is proved if we can show that this homomorphism is a reduction, since Theorem 5.1.8 and $F(\mathfrak{A}, G, v) \in \mathfrak{M}(\Sigma[I], \mathfrak{A})$ imply $F(\emptyset, G(I, \mathfrak{A}), v) \in \mathfrak{M}(\Sigma[I], \mathfrak{A})$.

We start with the surjectivity of $f\colon F(\emptyset, G(I, \mathfrak{A}), v) \to F(\mathfrak{A}, G, v)$. Let be $[t, G] \in F(\mathfrak{A}, G, v)_s$ and $s \in S$. This is equivalent to $\mathfrak{A} \vdash (v\colon G \to t \stackrel{e}{=} t)$, so that $(v\colon t \stackrel{e}{=} t) \in (v\colon G(I, \mathfrak{A}))$ and $\emptyset \vdash (v\colon G(I, \mathfrak{A}) \to t \stackrel{e}{=} t)$. This is equivalent to $[t, G(I, \mathfrak{A})] \in F(\emptyset, G(I, \mathfrak{A}), v)_s$, and so $[t, G] = [t, G(I, \mathfrak{A})]\, f_s$.

Next we prove the injectivity of $f_s: F(\emptyset, G(I, \mathfrak{A}), v)_s \to F(\mathfrak{A}, G, v)_s$ for every visible sort name $s \in I$.

Let be given $[t, G(I, \mathfrak{A})], [r, G(I, \mathfrak{A})] \in F(\emptyset, G(I, \mathfrak{A}), v)_s$ with

$$[t, G(I, \mathfrak{A})]\, f_s = [r, G(I, \mathfrak{A})]\, f_s,$$

so that $[t, G] = [r, G]$. This equation is equivalent to $\mathfrak{A} \vdash (v: G \to t \stackrel{\mathrm{e}}{=} r)$. Because of $t, r: v \to s$ and $s \in I$, it follows $(v: t \stackrel{\mathrm{e}}{=} r) \in (v: G(I, \mathfrak{A}))$ and therefore $\emptyset \vdash (v: G(I, \mathfrak{A}) \to t \stackrel{\mathrm{e}}{=} r)$, or in other terms $[t, G(I, \mathfrak{A})] = [r, G(I, \mathfrak{A})]$.

In a similar way we can prove the reflection property (3) of Definition 5.1.5. Let be $\sigma: s_1 \ldots s_n \to s$ any operator of $\Sigma[I]$ and

$$([t_1, G], \ldots, [t_n, G]) \in \mathrm{dom}(\sigma^{F(\mathfrak{A}, G, v)}).$$

Then we have

$$\mathfrak{A} \vdash \big(v: G \to \sigma(t_1, \ldots, t_n) \stackrel{\mathrm{e}}{=} \sigma(t_1, \ldots, t_n)\big),$$

so that $\big(v: \sigma(t_1, \ldots, t_n) \stackrel{\mathrm{e}}{=} \sigma(t_1, \ldots, t_n)\big) \in \big(v: G(I, \mathfrak{A})\big)$ and

$$\emptyset \vdash \big(v: G(I, \mathfrak{A}) \to \sigma(t_1, \ldots, t_n) \stackrel{\mathrm{e}}{=} \sigma(t_1, \ldots, t_n)\big).$$

The last statement implies

$$\big([t_1, G(I, \mathfrak{A})], \ldots, [t_n, G(I, \mathfrak{A})]\big) \in \mathrm{dom}(\sigma^{F(\emptyset, G(I, \mathfrak{A}), v)})$$

which completes the proof. ■

In the following we will always assume that the premises of conditional equations produce values in visible sorts. This condition, which is necessary for the proof of Theorem 5.1.12, is not a serious restriction with respect to applications. If one assumes that for each data type the behavioural equivalence is specified too, then one can easily transform an arbitrary existence equation into an equivalent one that produces values in the visible sort Bool.

Finally we prove that every $\Sigma[I]$-machine $M \in \mathfrak{M}(\Sigma[I], \mathfrak{A})$ is behaviourally equivalent to a $\Sigma[I]$-machine M^* that identically satisfies each elementary implication in $\mathfrak{A}$. This implies $\mathfrak{M}(\Sigma[I], \mathfrak{A}) \subseteq \mathfrak{M}^{\#}(\Sigma[I], \mathfrak{A})$ and gives the announced positive answer to the question arised above.

Theorem 5.1.13. *For every* I/0-*signature* $\Sigma[I]$, *every set* $\mathfrak{A}$ *of elementary implications, and every* $M \in \mathfrak{M}(\Sigma[I], \mathfrak{A})$ *there exists an I-equivalent* $\Sigma[I]$-*machine* M^* *satisfying each elementary implication in* $\mathfrak{A}$ *identically.*

Proof. In Section 3.3 it was shown that for every Σ-algebra A there is a set of existence equations $(v_A: G_A)$, with $v_A: X_A \to S$, so that A is isomorphic to $F(\emptyset, G_A, v_A)$.

There are several sets of existence equations representing A in this way. For the following proof we take any one of these. This is suitable, since every $\Sigma[I]$-machine A can be interpreted as a Σ-algebra. Now we consider the following $\Sigma[I]$-machines:

$$F(\emptyset, G, v) \in \mathfrak{M}(\Sigma[I], \mathfrak{A}),$$
$$F\big(\emptyset, G(I, \mathfrak{A}), v\big) \in \mathfrak{M}(\Sigma[I], \mathfrak{A}),$$
$$F(\mathfrak{A}, G, v) \in \mathfrak{M}(\Sigma[I], \mathfrak{A}),$$

where (v, G) is a set of existence equations such that M and $F(\emptyset, G, v)$ are isomorphic. Then there are two homomorphisms:

$$h\colon F\big(\emptyset, G(I, \mathfrak{A}), v\big) \to F(\emptyset, G, v),$$
$$g\colon F\big(\emptyset, G(I, \mathfrak{A}), v\big) \to F(\mathfrak{A}, G, v),$$

defined by

$$[t, G(I, \mathfrak{A})]_\emptyset\, h = [t, G]_\emptyset,$$
$$[t, G(I, \mathfrak{A})]_\emptyset\, g = [t, G]_\mathfrak{A}$$

where the indices $\emptyset$ and $\mathfrak{A}$ indicate the equivalence classes constructed with respect to the empty set $\emptyset$ of axioms and with respect to the set $\mathfrak{A}$ of axioms, respectively. It was proved above that $g\colon F\big(\emptyset, G(I, \mathfrak{A}), v\big) \to F(\mathfrak{A}, G, v)$ is a reduction. Theorem 5.1.13 is proved if we can show that the homomorphism

$$h\colon F\big(\emptyset, G(I, \mathfrak{A}), v\big) \to F(\emptyset, G, v)$$

is a reduction, too.

Surjectivity: Let $[t, G]_\emptyset$ be any element of $F(\emptyset, G, v)$. Then we have $\emptyset \vdash (v\colon G \to t \overset{e}{=} t)$ and $\mathfrak{A} \vdash (v\colon G \to t \overset{e}{=} t)$. This means $(v, t \overset{e}{=} t) \in \big(v\colon G(I, \mathfrak{A})\big)$, so that $\emptyset \vdash \big(v\colon G(I, \mathfrak{A}) \to t \overset{e}{=} t\big)$. But this proves $[t, G(I, \mathfrak{A}]_\emptyset\, h = [t, G]_\emptyset$.

Injectivity: Let $[t, G(I, \mathfrak{A})]_\emptyset$ and $[r, G(I, \mathfrak{A})]_\emptyset$ be elements of $F\big(\emptyset, G(I, \mathfrak{A}), v\big)_s$ with $s \in I$ and $[t, G(I, \mathfrak{A})]_\emptyset\, h_s = [r, G(I, \mathfrak{A})]_\emptyset\, h_s$. This implies $[t, G]_\emptyset = [r, G]_\emptyset$ which is equivalent to $\emptyset \vdash (v\colon G \to t \overset{e}{=} r)$. This latter relation implies $\mathfrak{A} \vdash (v\colon G \to t \overset{e}{=} r)$ so that $(v\colon t \overset{e}{=} r) \in \big(v\colon G(I, \mathfrak{A})\big)$ since $t, r\colon v \to s$ with $s \in I$. This finally yields $[t, G(I, \mathfrak{A})]_\emptyset = [r, G(I, \mathfrak{A})]_\emptyset$. Property (3) of Definition 5.1.5 can be proved in the same way as above. ■

We will illustrate the construction of a $\Sigma[I]$-machine $F(I, \mathfrak{A}, G, v)$ by the example of stacks. Therefore let the following hold:

$\Sigma[I]_{\text{STACK}}$ **is visible sorts** *Data*
  hidden sorts Stack
  oprn empty ⤳ Stack
    push(Stack,Data) ⤳ Stack
    top(Stack) ⤳ Data
    pop(Stack) ⤳ Stack

$$\mathfrak{A}_{\text{STACK}} = \{(\lambda\colon \emptyset \to \text{empty} = \text{empty}), \qquad \text{(E 1)}$$
$$(w\colon \emptyset \to \text{pop(push}(s, x)) = s), \qquad \text{(E 2)}$$
$$(w\colon \emptyset \to \text{top(push}(s, x)) = x)\} \qquad \text{(E 3)}$$

with $w\colon \{x, s\} \to \{\text{Data,Stack}\}$, $xw = \text{Data}$, $sw = \text{Stack}$, and $\lambda\colon \emptyset \to \{\text{Data, Stack}\}$. The set $(I, v\colon G)$ of I-equivalences is given by $v\colon D \to \{\text{Data,Stack}\}$ with $xv = \text{Data}$ for all $x \in D$, and $G = \emptyset$, i.e., we consider stacks $M(D)$ that are freely generated in $\mathfrak{M}(\Sigma[I]_{\text{STACK}}, \mathfrak{A}_{\text{STACK}})$ by an arbitrary set D of elements.

In such a stack $M(D)$ the constant operation 'empty$^{M(D)}$' is defined because of axiom (E 1), and it creates the empty stack.

Each one of the two axioms (E 2) and (E 3) makes the operation 'push$^{M(D)}$' everywhere defined, whereas the operations 'pop$^{M(D)}$' and 'top$^{M(D)}$' are executable if and only if 'push$^{M(D)}$' has been applied before at least one more than 'pop$^{M(D)}$'.

The set $M(D)_{\text{Stack}}$ consists of all Data-pruned terms built of elements of $M(D)_{\text{DATA}} = D$ and of the operators 'empty', 'push', and 'pop'.

Using the uniqueness of $M(D)$ up to isomorphisms, we give a more illustrative representation of the internal states of $M(D)$, i.e. of the elements of $M(D)_{\text{Stack}}$.

$M(D)_{\text{Stack}}$ = set of all finite trees, whose nodes are marked with elements of D, with the exception of the root, and which are equipped with a pointer at a node of the rightmost branch of the tree.

empty$^{M(D)}$ denotes the tree consisting only of the root.

The correspondence of these pointed D-marked trees to Data-pruned terms may be illustrated by the following examples:

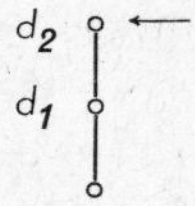

corresponds to

$\tau_1 = \text{push}(\text{push}(\text{empty}, d_1), d_2)$

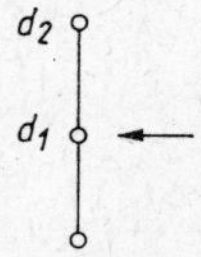

corresponds to

$\tau_2 = \text{pop}(\tau_1)$
$\quad = \text{pop}(\text{push})\text{push}(\text{empty}, d_1), d_2))$

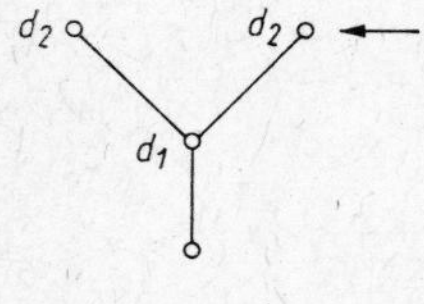

corresponds to

$\tau_3 = \text{push}(\tau_2, d_2)$
$\quad = \text{push}(\text{pop}(\text{push}(\text{push}(\text{empty}, d_1), d_2)), d_2)$

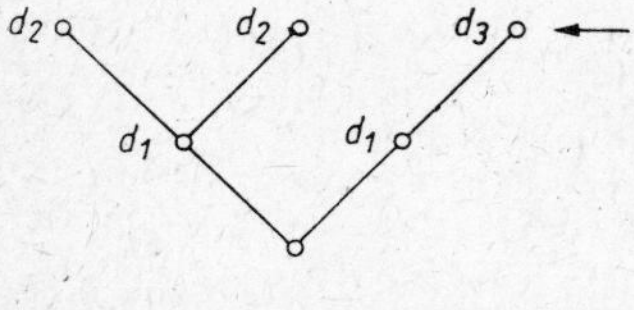

corresponds to

$\tau_4 = \text{push}(\text{push}(\text{pop}(\text{pop}(\tau_3)), d_1), d_3)$

These examples also demonstrate how the operations 'push$^{M(D)}$' and 'pop$^{M(D)}$' act on pointed D-marked trees.

Any two pointed D-marked trees are Data-equivalent if and only if their rightmost branches generates identical sequences of elements of D reading the marks from the root up to the pointer. Thus, τ_2 and pop(τ_4) are Data-equivalent and they are Data-equivalent to

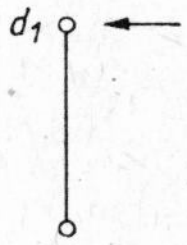

corresponding to
$\tau_5 = \text{push}(\text{empty}, d_1)$

The set of all those pointed D-marked trees which consists of only one branch with the pointer position at the leaf of that branch forms a set of representatives of the Data-equivalence. This set can be taken for the set of internal states of a STACK-machine of minimal redundancy. In our example the STACK-machine of minimal redundancy is that one, which is freely generated by D with respect to identical satisfaction of the same set of elementary implications. However, it is not generally true that the $\Sigma[I]$-machine $F(\mathfrak{A}, G, v)$, freely generated by a set $(v\colon G)$ of existence equations in ALG($\mathfrak{A}$), is the universal state minimalization of $F(I, \mathfrak{A}, G, v)$, since $F(\mathfrak{A}, G, v)$ might have distinct but Data-equivalent elements. In any case, $F(\mathfrak{A}, G, v)$ and $F(I, \mathfrak{A}, G, v)$ are I-equivalent $\Sigma[I]$-machines. We finish this section with the construction of a universal state minimalization for every $\Sigma[I]$-machine $F(I, \mathfrak{A}, G, v)$.

Theorem 5.1.14. *For every $\Sigma[I]$-machine $F(I, \mathfrak{A}, G, v)$ freely generated in $\mathfrak{M}(\Sigma[I], \mathfrak{A})$ by $(I, v\colon G)$ there exists an I-equivalent $\Sigma[I]$-machine $R(I, \mathfrak{A}, G, v)$ of minimal redundancy, i.e. any I-equivalent elements in $R(I, \mathfrak{A}, G, v)$ are equal.*

Proof. $R(I, \mathfrak{A}, G, v)$ can be constructed from $F(\mathfrak{A}, G, v)$ by identifying all I-equivalent elements. If $\boldsymbol{e} \in F(\mathfrak{A}, G, v)_{(v:G)}$ is the universal solution, see condition (2) of Definition 3.2.1, then we define

$$(v\colon G_I) = (v\colon G) \cup \{(v\colon t \overset{\mathrm{e}}{=} r) \mid \mathfrak{A} \vdash (v\colon G \to t \overset{\mathrm{e}}{=} t),\ \mathfrak{A} \vdash (v\colon G \to r \overset{\mathrm{e}}{=} r),$$
$$\boldsymbol{e}(t^{F(\mathfrak{A},G,v)}) \equiv \boldsymbol{e}(r^{F(\mathfrak{A},G,v)}) \bmod I\}.$$

Without any problems one can show that $R(I, \mathfrak{A}, G, v) = F(\mathfrak{A}, G_I, v)$ is of minimal redundancy. Since $F(\mathfrak{A}, G, v) \equiv F(\mathfrak{A}, G_I, v) \bmod I$ and $F(\mathfrak{A}, G, v) \equiv F(I, \mathfrak{A}, G, v) \bmod I$ the theorem is proved. ■

5.2. Initial restrictions of behaviour. Canons of behaviour

The existence of $\Sigma[I]$-machines $F(I, \mathfrak{A}, G, v)$ freely generated in $\mathfrak{M}(\Sigma[I], \mathfrak{A})$ by a set $(I, v\colon G)$ of I-equivalences makes possible a straightforward generalization of initial canons to canons of behaviour. What we basically have to do is to use behavioural satisfaction instead of identical satisfaction, and to exchange isomorphisms by reductions.

First we have to consider the consequences of behavioural semantics on the level of algebraic theories. The corresponding problems on the level of model classes have been considered in the previous section.

Definition 5.2.1. A *behavioural theory* $T = (\Sigma[I], \mathfrak{A})$ is given by an I/O-signature $\Sigma[I]$ and by a finite set $\mathfrak{A}$ of elementary implications. As introduced above,

$$\mathfrak{M}(\Sigma[I], \mathfrak{A})$$

denotes the class of all $\Sigma[I]$-machines that behaviourally satisfy each elementary implication in $\mathfrak{A}$.

A theory morphism

$$\varphi: (\Sigma_1[I_1], \mathfrak{A}_1) \to (\Sigma_2[I_2], \mathfrak{A}_2)$$

between behavioural theories is given by a signature morphism

$$\varphi = (\varphi_{\text{sorts}}, \varphi_{\text{operators}}): \Sigma_1 \to \Sigma_2$$

(see Definition 2.2.2) satisfying:

(1) $A \models \text{mod}\, I_2(v: G \to t \stackrel{e}{=} r)\,\varphi$ holds for every $A \in \mathfrak{M}(\Sigma_2[I_2], \mathfrak{A}_2)$ and every elementary implication $(v: G \to t \stackrel{e}{=} r) \in \mathfrak{A}_1$;

(2) $\varphi_{\text{sorts}}: S_1 \to S_2$ preserves visible sorts, i.e. $s \in I_1$ implies $s\varphi_{\text{sorts}} \in I_2$. ■

As distinct from the corresponding definition of a specification morphism by Goguen and Meseguer (1982) we require the sort function to preserve visible sorts, whereas in that paper the preservation of hidden sorts is required. If we follow the last requirement, then we can not prove that for every reduction $r: A \to B$ of $\Sigma_2[I_2]$-machines the induced $\Sigma_1[I_1]$-homomorphism

$$r\dot{\varphi}: A\varphi \to B\varphi$$

is a reduction, too. The preservation of reductions by forgetful functors is a very natural requirement, since reductions play in the context of behavioural semantics a role which is comparable with that of isomorphism in the context of identical satisfaction of elementary implications. On the other hand, if one takes into account the complete paper, than one can be sure that the required preservation of hidden sorts is a misprint.

There is no doubt that the results of Section 3.5 concerning the existence of relatively free partial algebras can be extended to behavioural theories. We omit here the details of that extension and turn to the extension of Chapter 4.

Definition 5.2.2. Let the following be given: behavioural theories $T = (\Sigma[I], \mathfrak{A})$, $T_2 = (\Sigma_2[I_2], \mathfrak{A}_2)$, a behavioural subtheory $T_1 = (\Sigma_1[I_1], \mathfrak{A}_1)$ fo T_2, and a theory morphism $\varphi: T_2 \to T$. The ordered pair

$$(T_1, \varphi: T_2 \to T)$$

is called an *initial restriction of behaviour in T*.

A T-machine M, i.e. $M \in \mathfrak{M}(T)$, satisfies this initial restriction of behaviour, in symbols

$$M \models (T_1, \varphi: T_2 \to T),$$

if:

(1) $F{\downarrow}T_1 = (M\varphi){\downarrow}T_1$, where $F \in \mathfrak{M}(T_2)$ is freely generated by $(M\varphi){\downarrow}T_1$ according to $T_1 \subseteq T_2$;

(2) The homomorphism $f: F \to M\varphi$, uniquely determined by $f{\downarrow}T_1 = \mathrm{id}\big((M\varphi){\downarrow}T_1\big)$, is a reduction. ■

As distinct from Definition 4.1.1 it is now not required that M itself is freely generated by $(M\varphi){\downarrow}T_1$ in $\mathfrak{M}(T_2)$. This requirement would yield once more isomorphism classes as semantical counterparts of data abstraction. Since reductions coincide with isomorphisms if all sort names are visible, Definition 4.1.1 is included in the previous one.

In the case of identical satisfaction the condition $F{\downarrow}T_1 = (M\varphi){\downarrow}T_1$ is not explicitly required, because it is an implication of the requirement that $M\varphi$ is freely generated by $(M\varphi){\downarrow}T_1$. However, in the case of behavioural semantics we have explicitly to state the *strong persistency of an initial enrichment* and of a machine satisfying the initial restriction of behaviour. If we omitted condition (1), we could only prove that $(M\varphi){\downarrow}T_1$ and $(M'\varphi){\downarrow}T_1$ were I_1-equivalent if M and M' satisfied the initial restriction of behaviour $(T_1, \varphi: T_2 \to T)$. In the next section we will see that condition (1) is very significant for the concept of initial computability in canons of behaviour.

Notice, condition (2) in Definition 5.2.2 does not imply that a T-machine M' which is I-equivalent to a T-machine M with $M \models (T_1, \varphi: T_2 \to T)$ itself satisfies that initial restriction of behaviour in T, not even if additionally $(M'\varphi){\downarrow}T_1 = (M\varphi){\downarrow}T_1$. Condition (2) implies that the sets of internal states in $M\varphi$ are reachable with respect to $(M\varphi){\downarrow}T_1$.

Definition 5.2.3. *A canon of behaviour* (or *behavioural canon*)

$$\mathbb{C} = (T, \Delta)$$

is given by a behavioural theory T and by a finite set Δ of initial restriction of behaviour in T.

$\mathfrak{M}(\mathbb{C})$ and $\mathfrak{M}(T, \Delta)$ denote the class of all T-machines satisfying each initial restriction of behaviour in Δ.

A morphism

$$\varphi: (T_1, \Delta_1) \to (T_2, \Delta_2)$$

between canons of behaviour is given by a theory morphism $\varphi: T_1 \to T_2$ so that $A \in \mathfrak{M}(T_2, \Delta_2)$ implies $A\varphi \in \mathfrak{M}(T_1, \Delta_1)$. ■

In textual representations of canons of behaviour we have only to make visible the division of sort names in visible and hidden sort names. We will use the convention that a sort name is visible if it is not explicitly indicated as a hidden sort name. According to this convention a *behavioural specification* of stacks would look like:

STACK is DATA **with definition**
  hidden sort Stack
  oprn empty $\rightarrow$ Stack
    push(Stack,Data) $\rightarrow$ Stack
    top(Stack) $\circ\!\!\rightarrow$ Data
    pop(Stack) $\circ\!\!\rightarrow$ Stack
  axioms s: Stack, d: Data
    top(push(s,d)) $= d$
    pop(push(s,d)) $= s$
end STACK

A behavioural specification of sequences looks just like the specification in the introduction of this chapter, since according to generally accepted understanding both 'Data' and 'Seq' are visible sort names.

Now we are in a good position, on flashing back to the renaming of the specifications STACK and SEQUENCE given in the introduction of this chapter, since we are able to answer the question of whether this renaming describes a canon morphism

$$\varphi\colon \mathfrak{C}_{\text{SEQUENCE}} \to \mathfrak{C}_{\text{STACK}}.$$

Checking Definition 5.2.1 we see that the renaming maps 'Seq' onto 'Stack', and so it maps a visible sort onto a hidden sort. Hence, that renaming does not describe a canon morphism. However, the inverse renaming

$$\varphi\colon \text{Stack} \mapsto \text{Seq, Data} \mapsto \text{Data, empty} \mapsto \text{nil, push} \mapsto \text{cons,}$$
$$\text{top} \mapsto \text{head, pop} \mapsto \text{tail}$$

describes a canon morphism

$$\varphi\colon \mathfrak{C}_{\text{STACK}} \to \mathfrak{C}_{\text{SEQUENCE}}.$$

The canon morphism $\varphi\colon \mathfrak{C}_{\text{STACK}} \to \mathfrak{C}_{\text{SEQUENCE}}$ can be interpreted as an implementation of stacks by sequences. However, it is impossible to implement sequences by stacks, since there is no canon morphism in this direction as far as stacks are specified as a hidden data type. This asymmetry makes clearly visible the serious differences between both these data types, which could not be described without behavioural semantics.

Next we extend the notion of a persistent canon morphism.

Definition 5.2.4. A behavioural canon morphism

$$\varphi\colon (T_1, \Delta_1) \to (T_2, \Delta_2)$$

is said to be *persistent*, if:

(1) For every $B \in \mathfrak{M}(T_1, \Delta_1)$ there is an $F \in \mathfrak{M}(T_2, \Delta_2)$ freely generated by B according to $\varphi: T_1 \to T_2$ such that $F\varphi = B$.

(2) If $F, A \in \mathfrak{M}(T_2, \Delta_2)$, if F is freely generated by $A\varphi$ with $F\varphi = A\varphi$, then the uniquely determined homomorphism $f: F \to A$ with $f\varphi = \mathrm{id}(A\varphi)$ is a reduction. ■

One can use just the same proof as in Section 4.3 to show that pushouts exist for behavioural canon morphisms, to prove that behavioural subcanons are preserved by pushouts, and that statements (1) and (2) of Theorem 4.3.4 are also true for the more general concept of behavioural canon morphisms.

The proof of the extended version of statement (4) in Theorem 4.3.4, saying that pushouts preserve persistency, requires a bit more, and will be given in the following.

Corollary 5.2.5. *If the diagram*

$$\begin{array}{ccc} (T_0, \Delta_0) & \xrightarrow{\ \varepsilon\ } & (T_1, \Delta_1) \\ {\scriptstyle\varphi}\downarrow & & \downarrow{\scriptstyle\varphi^{\#}} \\ (T_2, \Delta_2) & \xrightarrow{\ \varepsilon^{\#}\ } & (T, \Delta) \end{array}$$

is a pushout of behavioural canon morphisms, and if $\varepsilon: (T_0, \Delta_0) \to (T_1, \Delta_1)$ *is persistent, then* $\varepsilon^{\#}: (T_2, \Delta_2) \to (T, \Delta)$ *is persistent too.*

Proof: For the proof of condition (1) we can take the corresponding proof of statement (4) in Theorem 4.3.4.

For the proof of condition (2) let $F, A \in \mathfrak{M}(T, \Delta)$, F be freely generated by $A\varepsilon^{\#}$ with $A\varepsilon^{\#} = F\varepsilon^{\#}$, and $f: F \to A$ be the uniquely determined homomorphism with $f\varepsilon^{\#} = \mathrm{id}(A^{\#})$. It remains to prove that this homomorphism is a reduction.

Because of the pushout construction $F \in \mathfrak{M}(T, \Delta)$ implies $(F\varepsilon^{\#})\,\varphi = (F\varphi^{\#})\,\varepsilon$. Since $\varepsilon: (T_0, \Delta_0) \to (T_1, \Delta_1)$ is persistent, there is a machine $H \in \mathfrak{M}(T_1, \Delta_1)$ freely generated by $(F\varphi^{\#})\,\varepsilon$ with $H\varepsilon = (F\varphi^{\#})\,\varepsilon$, and there is a reduction $h: H \to F\varphi^{\#}$ with $h\varepsilon = \mathrm{id}\big((F\varphi^{\#})\,\varepsilon\big)$. Because of $(A\varphi^{\#})\,\varepsilon = (A\varepsilon^{\#})\,\varphi = (F\varepsilon^{\#})\,\varphi = (F\varphi^{\#})\,\varepsilon$ there is a reduction $r: H \to A\varphi^{\#}$ with $r\varepsilon = \mathrm{id}\big((F\varphi^{\#})\,\varepsilon\big)$. The uniqueness of these both reductions and $\big(h \circ (f\varphi^{\#})\big)\,\varepsilon = (h\varepsilon) \circ (f\varphi^{\#})\,\varepsilon = (h\varepsilon) \circ \big((f\varepsilon^{\#})\,\varphi\big) = h\varepsilon$, because of $f\varepsilon^{\#} = \mathrm{id}(A\varepsilon^{\#})$, implies

$$r = h \circ (f\varphi^{\#}),$$

and this equation makes $f\varphi^{\#}$ a reduction. Now we know that $f\varphi^{\#}: F\varphi^{\#} \to A\varphi^{\#}$, and $f\varepsilon^{\#} = \mathrm{id}(A\varepsilon^{\#})$ are reductions, so that $f = f\varepsilon^{\#} \oplus f\varphi^{\#}$ is a reduction, too. ■

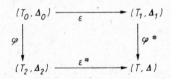

Finally we will make the specification of stacks a bit more realistic, giving a behavioural specification of bounded stacks. In the following section we describe an implementation of bounded stacks by pointed arrays by means of a suitable morphism between canons of behaviour.

B-STACK is DATA **with** NAT **with definition**
  hidden sorts Stack
  oprn empty(Nat) $\to$ Stack
    limit(Stack) $\to$ Nat
    depth(Stack) $\to$ Nat
    push(Stack,Data) $\circ\!\!\to$ Stack
    top(Stack) $\circ\!\!\to$ Data
    pop(Stack) $\circ\!\!\to$ Stack
  axioms m: Nat, s: Stack, d: Data
    limit(empty(m)) $= m$
    depth(empty(m)) $=$ zero
    if depth(s) $<$ limit(s) **then** depth(push(s,d)) $=$ succ(depth(s))
    if depth(s) $<$ limit(s) **then** limit(push(s,d)) $=$ limit(s)
    if depth(s) $<$ limit(s) **then** top(push(s,d)) $= d$
    if depth(s) $<$ limit(s) **then** pop(push(s,d)) $= s$
end B-STACK

In this specification 'empty(m)' creates a bounded, empty stack with a maximal depth of m. Therefore the 'push'-operation becomes also partial, and it is applicable if and only if the depth is less than the maximal depth.

5.3. Initial computability in canons of behaviour

In this section we extend the concept of initial computability in canons given by Definition 4.3.

Now we have to take into account the existence of hidden sort names and of visible sort names. With respect to visible sort names nothing has been changed. What about hidden sort names? The construction of machines of type $F(I, \mathfrak{A}, G, v)$ indicates a first essential difference between visible data types and hidden data types. The latter would be better called modules. This difference concerns prototyping. A visible data type has a computable prototype if and only if its identity relation can be specified by a definition enrichment. A machine of maximal redundancy has always a computable prototype providing we are able to decide if any two I-terms represent equal input values.

More formally we can say that terms $t', t'' : v \to s$, with $v: X \to S$ and $s \notin I$ represent the same internal state in $F(I, \mathfrak{A}, G, v)_s$, if there is a term $t: w \to s$ with $w: X_0 \to S$, $X_0 = \{x \in X \mid xv \in I\}$ and $xw = xv$ for all $x \in X_0$, and if there are assignments $\boldsymbol{r}', \boldsymbol{r}'' \in T(\Sigma, v)_w$ with $t' = t\tilde{\boldsymbol{r}}'$, $t'' = t\tilde{\boldsymbol{r}}''$ and

$\boldsymbol{r}'$, $\boldsymbol{r}''$ represent pointwise I-equivalent input values. The term t is then the common I-pruned term of t' and t''. This observation implies that $F(I, \mathfrak{A}, G, v)$ has a computable s-carrier if and only if all possible input types of s have computable prototypes. This fact justifies the following definition.

Definition 5.3.1. Let $\mathbf{C} = ((\Sigma[\mathrm{I}], \mathfrak{A}), \Delta)$ be a canon of behaviour and $\mathbf{C}_0$ any subcanon of $\mathbf{C}$.
A sort name $s' \in I$ is called an *input sort* of the sort name $s \notin I$, if there is a term $t: v \to s$, with $v: X \to S$, and if there is a variable $x_0 \in X$ which occurs in t and which is of sort s', i.e. $x_0 v = s'$.

A hidden sort name s of $\mathbf{C}$ is called *weekly recursive over* $\mathbf{C}_0$, if each input sort of s is recursive over $\mathbf{C}_0$. ■

The definition of a sort name which is recursive over a subcanon can also be extended to hidden sort names of a canon of behaviour. But now we have to require that the I-equivalence of elements can be specified by a definition enrichment, i.e., in Definition 4.3.2 we have to require that

$$(s^*\text{-eq})^B: A_s \times A_s \to \{\text{true, false}\}$$

is the relation of I-equivalence. We omit here a repetition of the definition with that alteration in its wording.

If a hidden sort name is recursive over a subcanon, then this is equivalent to the statement that there is a computable prototype of minimal redundancy providing all sort names of the subcanon have a computable prototype.

We will now consider functional enrichments of canons of behaviour, or equivalently operators that are recursive over a subcanon. Looking at the specification SEQUENCE we observe that this enrichment by definition can be split up into three successive enrichments by definition:

```
SEQUENCE is DATA with definition
     sorts Seq
     oprn nil → Seq
          cons(Seq,Data) → Seq
     with definition
     oprn head(Seq) ⤞ Data
     axioms s: Seq, d: Data
          head(cons(s,d)) = d
     with definition
     oprn tail(Seq) ⤞ Seq
     axioms s: Seq, d: Data
          tail(cons(s,d)) = s
end SEQUENCE
```

The splitting up into three successive enricments by definitions represents the fact that a minimal set of operations necessary for the generation of all sequences is given by {nil, cons}. In many papers these operations are called the 'generators' of the data type. The other two enrichments by definition do not define a new data type. The purpose of this enrichments is the enrichment of the functionality of the previously defined data type. In our terminology this is described by the fact that the operators 'head' and 'tail' are recursive over the first enrichment by definition. We claim that enriching the functionality of a data type does not change the meaning of that data type, it only improves its usefulness.

If we look at the specification STACK and ask if all the given operations are necessary for the specification of stacks, or if one of them is recursive over the remainder, then we will see another essential different between hidden sort names and visible ones.

A first look says that stacks cannot be specified using only 'empty' and 'push'. The lack of any output operation would imply a collapsing, i.e. any two elements of sort 'Stack' are I-equivalent. Hence we need a least 'empty', 'push', and 'top'. The question arises if 'pop' is recursive over these three operators.

Let us attempt a specification of stacks without the 'pop'-operation, and let us try a specification of the 'pop'-operation by a second enrichment by definition.

STACK-1 **is** DATA **with definition**
    hidden sorts Stack
    oprn empty $\rightarrow$ Stack
        top(Stack) $\circ\!\!\rightarrow$ Data
        push(Stack,Data) $\rightarrow$ Stack
    axioms s: Stack, d: Data
        top(push(s,d)) $= d$
end STACK-1

In this canon of behaviour the machines of minimal redundancy and of maximal redundancy coincide, since the machine freely generated by an arbitrary set of data is of minimal redundancy.

Now we attempt a specification of the 'pop'-operation:

STACK-2 **is** STACK 1 **with definition**
    oprn pop(Stack) $\circ\!\!\rightarrow$ Stack
    axioms s: Stack, d: Data
        pop(push(s,d)) $= s$
end STACK-2

This enrichment by definition is not a persistent enlargement of STACK-1. To demonstrate this let $\Sigma_1[I_1]$ be the I/O-signature of STACK-1, $\Sigma_2[I_2]$

the I/O-signature of STACK-2 and $\mathfrak{A}_2$ the set of elementary implications of STACK-2. If $M \in \mathfrak{M}(\mathfrak{C}_{\mathrm{STACK-1}})$, and if F is freely generated by M in $\mathfrak{M}(\Sigma_2[I_2], \mathfrak{A}_2)$, then F is isomorphic to the $\Sigma_2[I_2]$-machine $M(D)$ defined in Section 5.1. But this implies

$$F\downarrow(\Sigma_1[I_1], \mathfrak{A}_1) \notin \mathfrak{M}(\mathfrak{C}_{\mathrm{STACK-1}})$$

since $F\downarrow(\Sigma_1[I_1], \mathfrak{A}_1)$ has additional unreachable internal states produced by mixed applications of 'pop' and 'push'. Thus condition (1) of Definition 5.2.2 is not satisfied for all machines of the kind $M(D)$ if $D \neq \emptyset$, and so $\mathfrak{M}(\mathfrak{C}_{\mathrm{STACK-2}})$ has up to isomorphisms only one element $M(\emptyset)$, which is the empty stack with the empty set $D = \emptyset$ of storable values.

This consideration shows that a behavioural specification of stacks requires all four operators 'empty', 'push', 'top', and 'pop', and in addition it is impossible to define a new operator that produces elements of sort 'Stack' by any definition enrichment. This effect does not depend on stacks, it is representative for the general situation.

Theorem 5.3.2. *If $\mathfrak{C} \subseteq \mathfrak{C}'$ is a persistent enlargement of canons of behaviour without additional hidden sort names in $\mathfrak{C}'$, and if $\sigma: w \to s$ is an operator in $\mathfrak{C}'$ not contained in $\mathfrak{C}$, then the sort name s is a visible sort name in $\mathfrak{C}'$.*

This result can be considered in some sense as a justification or as a refutation of Definition 5.2.2 depending on how one understands intuitively computability on parameterized program modules. We do not understand this section as a definite position. On the contrary, we interpret it as a serious sign of an insufficient understanding of the concept of computable enrichments of the functionality of program modules. Having in mind the concept of *Parameterized Programming,* suggested by GOGUEN and MESEGUER (1984), we believe that the understanding of those functional enrichments should be well-founded. We strongly support the claim of GOGUEN and MESEGUER that parameterized programming, as a form of programming-in-the-large, permits an unusually high degree of reusability.

We finish with the demonstration that the well-known realization of bounded stacks by pointed arrays can be described as a behavioural canon morphism. We begin with the specification of one-dimensional arrays.

ARRAY is DATA **with** NAT
  with requirement
  oprn $\delta \to$ Data
  with definition
  sorts Array

```
    oprn create(Nat) → Array
         bound(Array) → Nat
         write(Array,Nat,Data) ↛ Array
         read(Array,Nat) ↛ Data
    axioms n,i,j: Nat, x: Data, a: Array
         bound(create(n)) = n
         if 1 ≤ i < bound(a) then bound(write(a,i,x)) = bound(a)
         if 1 ≤ i ≤ n then read(create(n),i) = δ
         if 1 ≤ i,j ≤ bound(a), i ≠ j then
              read(write(a,i,x),j) = read(a,j) fi
         if 1 ≤ i ≤ bound(a) then read(write(a,i,x),i) = x
end ARRAY
```

An implementation of bounded stacks by pointed arrays is a typical example since we have to start with a definition enrichment of ARRAY which specifies a new data type 'pointed arrays', represented by the sort name 'P-Array', and which implements the stack operations by a definition of appropriate operations on pointed arrays.

```
P-ARRAY is ARRAY with definition
    sorts P-Array
    oprn pair(Nat, Array) ↛ P-Array
    axioms n: Nat, a: Array
         if n ≤ bound(a) then pair(n,a) = pair(n,a)
    with definition
    oprn empty*(Nat) → P-Array
         limit*(P-Array) → Nat
         depth*(P-Array) → Nat
         push*(P-Array,Data) ↛ P-Array
         top*(P-Array) ↛ Data
         pop*(P-Array) ↛ P-Array
    axioms n: Nat, a: Array, x: Data
         empty*(n) = pair(zero,create(n))
         depth*(pair(zero,create(zero))) = zero
         limit*(pair(zero,create(zero))) = zero
         if n ≤ bound(a) then limit*(pair(n,a)) = bound(a)
         if n ≤ bound(a) then depth*(pair(n,a)) = n
         if n < bound(a) then
              push*(pair(n,a),x) = pair(n +1,write(a,n+1,x)) fi
         if 1 ≤ n ≤ bound(a) then
              top*(pair(n,a)) = read(a,n) fi
         if 1 ≤ n ≤ bound(a) then
              pop*(pair(n,a)) = pair(n−1,a) fi
end P-ARRAY
```

The mappings

$$\varphi_{\text{sorts}}\colon \text{Data} \mapsto \text{Data}, \text{Stack} \mapsto \text{P-Array}$$

$$\varphi_{\text{operators}}\colon \text{empty} \mapsto \text{empty*}, \ldots, \text{pop} \mapsto \text{pop*}$$

represent a behavioural canon morphism

$$\varphi\colon \mathbb{C}_{\text{B-STACK}} \to \mathbb{C}_{\text{P-ARRAY}}$$

Evidently, the mappings φ_{sorts}, $\varphi_{\text{operators}}$ would not represent a behavioural canon morphism if the sort name 'Stack' in $\mathbb{C}_{\text{B-STACK}}$ were be a visible sort name.

References

ADJ (1975)

GOGUEN, J. A., THATHER, J. W., WAGNER, E. G., WRIGHT, J. B. (1975) *Initial algebraic semantics.* IBM Research Report RC 5364.

ADJ (1978)

GOGUEN, J. A., THATCHER, J. W., WAGNER, E. G. (1978) *An initial algebra approach to the specification, correctness, and implementation of abstract data types.* In: R. T. YEH (ed.), Current Trends in Programming Methodology, Vol. 4: Data Structuring, Prentice-Hall, 80—149.

ANDREKA, H., BURMEISTER, P., NEMETI, I. (1980) *Quasivarieties of partial algebras — unifying approach towards a two-valued model theory for partial algebras.* TH Darmstadt, FB Mathematik, Preprint Nr. 557.

BENECKE, K. (1979) *Signaturketten und Operations- und Mengenformen über Signaturketten.* Thesis, Technische Hochschule „Otto von Guericke" Magdeburg.

BENECKE, K. (1983) Spezifikation parametrischer Datentypen. *Zeitschr. f. math. Logik und Grundlagen d. Math.* **29**, 83—96.

BERGSTRA, J. A., BROY, M., TUCKER, J. V., WIRSING, M. (1981) On the power of algebraic specifications. In: GRUSKA, J. (ed.), 10th MFCS, Strbske Pleso, CSSR, Sept. 1981, LNCS 118, Springer-Verlag, Berlin—Heidelberg—New York, 193—204.

BERGSTRA, J. A., MEYER, J. (1981) I/0-computable data structures. *SIGPLAN Notices* **16**, 4, 27—32.

BERGSTRA, J. A., MEYER, J. (1983) On specifying sets of integers. *Journal of Information Processing and Cybernetics — EIK* **20**, 10—11, 531—541.

BROY, M., WIRSING, M. (1981) On the algebraic extension of abstract data types. In: DIAZ, J., RAMOS, I. (eds.), *Formalization of Programming Concepts*, LNCS 107, Springer-Verlag, Berlin—Heidelberg—New York, 244—251.

BROY, M., WIRSING, M. (1982) Partial Abstract Types. *Acta Informatica*, **18**, 47—64.

BIRKHOFF, G., LIPSON, J. D. (1970) Heterogeneous algebras. *J. Combinatorial Theory* 8, 115—133.

BURMEISTER, P. (1986) *A model theoretic oriented approach to partial algebras.* Mathematical Research 31, Akademie-Verlag, Berlin 1986.

BURSTALL, R. M., GOGUEN, J. A. (1977) Putting theories together to make specifications. In: *Proc. of 5th Int. Joint Conf. on Artifical Intelligence*, MIT, Cambridge, Mass., 1045—1058.

BURSTALL, R. M., GOGUEN, J. A. (1980) The semantics of Clear, a specification language. In: *Proc. of Advanced Course on Abstract Software Specifications*, Copenhagen. LNCS, 86, Springer-Verlag, Berlin—Heidelberg—New York, 292—332.

BURSTALL, R. M., GOGUEN, J. A. (1982) Algebras, Theories and Freeness: An Introduction for Computer Scientists. In: *Proceedings 1981 Marktoberdorf NATO Summer School*, Reidel, Dordrecht.

BURSTALL, R. M., MACQUEEN, D., SANNELLA, D. (1980) *HOPE: An Experimental Applicative Language.* Report CSR-62-80, Computer Science Dept., Edinburgh University.

EHRIG, H. (1981) Parameterized specification with requirements. *Proc. CAAP '81*, LNCS, 112, Springer-Verlag, Berlin—Heidelberg—New York, 1—24.

Ehrig, H., Thatcher, J. W., Lucas, P., Zilles, S. N. (1982) *Denotational and initial algebra semantics of the algebraic specification language LOOK*. Draft report, IBM research.

Ehrig, H., Wagner, E. G., Thatcher, J. W. (1983) Algebraic Specifications with Generating Constraints. In: *Proceedings ICALP 1983*, Barcelona, LNCS 154, Springer-Verlag, Berlin—Heidelberg—New York, 188—202.

Ganzinger, H. (1983) Parameterized specifications: parameter passing and implemention. *TOPLAS* **3**, 318—354.

Giarratana, V., Gimona, F., Montanari, U. (1976) Observability concepts in abstract data type specification. In: A. Mazurkiewicz (ed.), *MFCS*, 76, LNCS 45, Springer-Verlag, Berlin—Heidelberg—New York, 576—587.

Goguen, J. A. (1978) Abstract errors for abstract data types. *Proc. IFIP Working Conf. on the Formal Description of Programming Concepts*, MIT, 21.1—21.32.

Goguen, J. A. (1984) Parameterized Programming. *IEE Trans. of Softw. Engin.*, **10**, 528—543.

Goguen, J. A. (1984) Suggestions for using organizing libraries for Ada program development. *Techn. Rep. for Ada Joint Program Office*, SRI International.

Goguen, J. A. (1985) *Order sorted algebras*. Dep. Comput. Sci. Univ. California, Los Angeles, Semantics and Theory of Computation, Rep. 14.

Goguen, J. A., Burstall, R. M. (1980) *CAT, a system for the structured elaboration of correct programs from structured specifications*. Technical report CSL-118, Computer Science Laboratory, SRI International.

Goguen, J. A., Burstall, R. M. (1983) *Institutions: Abstract model theory for program specification*. Draft report, SRI International.

Goguen, J. A., Burstall, R. M. (1984) Introducing Institutions. In: Clarke, E., Kozen, D. (eds.), *Proc. Logics of Programming Workshop*, Springer-Verlag, Berlin—Heidelberg—New York, 221—256.

Goguen, J. A., Meseguer, J. (1981) Completeness of many-sorted equational logic. *SIGPLAN Notices* **16**, 7, 24—32.

Goguen, J. A., Meseguer, J. (1982) Universal Realization, Persistent Interconnection and Implementation of Abstract Modules, In: Nielsen, M., Schmidt, E. M. (eds.), *Proc. 9th ICALP*, LNCS 140, Springer-Verlag, Berlin—Heidelberg—New York, 265—281.

Grätzer, G. (1979) *Universal algebra*. 2nd ed., Springer-Verlag, New York—Heidelberg—Berlin.

Hoare, C. A. R. (1972) Proof of correctness of data representations. *Acta Informatica* **4**, 271—281.

Höft, H. (1973) Weak and strong equations in partial algebras. *Algebra Universalis* **3**, 203—215.

Hornung, G., Raulefs, P. (1980) Terminal algebra semantics and retractions for abstract data types. In: DeBakker, J., van Leeuwen, J. (eds.), *7th ICALP* (1980), LNCS 85, Springer-Verlag, Berlin—Heidelberg—New York, 310—323.

Hupbach, U. L. (1980) Abstract implementation of abstract data types. In: Dembiński, P. (ed.), *MFCS '80*, LNCS 88, Springer-Verlag, Berlin—Heidelberg—New York, 291—304.

Hupbach, U. L. (1981) Abstract Implementation and Parameter substitution. In: Arato, M., Varga, L. (eds.), *Proc. 3rd Hungarian Computer Science Conference*, Budapest.

Hupbach, U. L., Kaphengst, H., Reichel, H. (1980) *Initial algebraic specification of data types, parameterized data types, and algorithms*. VEB Robotron, Zentrum für Forschung und Technik, Dresden, WIB 15.

Kamin, S. (1983) Final data types and their specification. *ACM TOPLAS* **5** (1), 97–123.

Kamin, S., Archer, M. (1984) Partial Implementations of Abstract Data Types: A Dissenting View on Errors. In: Kahn, G., MacQueen, D. B., Plotkin, G. (eds.), *Proc. Int. Symposium on Semantics of Data Types*, Sophia-Antipolis, LNCS 173, Springer-Verlag, Berlin–Heidelberg–New York, 317 to 336.

Kaphengst, H. (1981) What is computable for abstract data types? In: Gécseg, F. (ed.), *Proc. FCT '81*, LNCS 117, Springer-Verlag, Berlin–Heidelberg–New York, 173–183.

Kaphengst, H., Reichel, H. (1971) *Algebraische Algorithmentheorie*. VEB Robotron, Zentrum für Forschung und Technik, Dresden.

Kerkhoff, R. (1970) Gleichungsdefinierbare Klassen partieller Algebren. *Math. Annalen* **185**, 112–133.

Knuth, D. E. (1968) Semantics of context-free languages. *Math. Syst. Theory* **2**, 127–145.

Letitshevski, A. A. (1968) Syntax and semantics of formal languages (Russian), *Cybernetics* **4**, 1–9.

Lugowski, H. (1976) *Grundzüge der Universellen Algebra*. Teubner-Texte zur Mathematik, Teubner-Verlag, Leipzig.

MacLane, S. (1971) *Categories for the Working Mathematician*. Springer-Verlag, Berlin–Heidelberg–New York.

Milner, R. (1978) A Theory of Type Polymorphism in Programming. *Journal of Computer and System Science* **17**, 3, 348–375.

Milner, R. (1984) A Proposal for Standard ML. *Report CSR*-157-83, Computer Science Dept., University of Edinburgh.

Poythress, V. S. (1979) Partial morphisms on partial algebras, *Algebra Universalis* **3**, 182–202.

Reichel, H. (1969) Kanonische Zerlegung von Funktoren. *Math. Nachr.* **41**, 4–6, 361–370.

Reichel, H. (1979) Theorie der Äquoide. Dissertation B, Humboldt Universität zu Berlin.

Reichel, H. (1980) Initially restricting algerbraic theories. In: Dembiński, P. (ed.), *Proc. 9th MFCS*, Ryszyna, LNCS 88, Springer-Verlag, Berlin–Heidelberg–New York, 504–514.

Reichel, H. (1981) Behavioural equivalence — a unifying concept for initial and final specification methods. In: Arato, M., Varga, L. (eds.), *Proc. 3rd Hungarian Computer Science Conference*, Budapest, 27–39.

Reichel, H. (1984) Behavioural Validity of conditional equations in abstract data types. In: *Contributions to General Algebra 3, Proc. of the Vienna Conference, June 21–24, 1984*, Verlag Hölder-Pichler-Tempsky, Wien 1985 — Verlag B. G. Teubner, Stuttgart, 301–324.

Reichel, H. (1985) Initial Restrictions of Behaviour. In: *Proc. IFIP Working Conference The Role of Abstract Models in Information Processing*, Vienna, January 1985.

ROGERS, H. (1967) *Theory of Recursive Functions and Effective Computability.* McGraw-Hill, New York.

SANNELLA, D., WIRSING, M. (1983) A kernel language for algebraic specification and implementation. *Report CSR-131-83,* Computer Science Dept., University Edinburgh.

SANNELLA, D. T., TARLECKI, A. (1984) Building specifications in an arbitrary institution. In: *Proc. Int. Symposium on Semantics of Data Types,* Sophia-Antipolis, LNCS 173, Springer-Verlag, Berlin—Heidelberg—New York, 337—356.

SANNELLA, D. T., TARLECKI, A. (1985) Program specification and development in Standard ML. In: *Proc. 12th ACM Symp. on Principles of Programming Languages,* New Orleans, January 1985.

SANNELLA, D. T., TARLECKI, A. (1985) On Observational Equivalence and Algebraic Specification. In: *Formal Methods and Software Development,* TAPSOFT Proc., LNCS 185, Springer-Verlag, Berlin—Heidelberg—New York, 308—322.

SCHMIDT, J. (1966) A general existence theorem on partial algebras and its special cases. *Coll. Math.* **14**, 73—87.

SCOTT, D. (1981) *Lectures on a Mathematical Theory of Computation.* Oxford University Computing Laboratory, Technical Monograph PRG-19.

SHOENFIELD, J. R. (1967) *Mathematical Logic.* Addision-Wesley, Reading (Mass.).

SLOMINSKI, J. (1968) Peano-algebras and quasi-algebras. Dissertationes Mathematicae 62.

THIELE, H. (1966) *Wissenschaftstheoretische Untersuchungen in algorithmischen Sprachen I (Theorie der Graphschemata-Kalküle).* VEB Deutscher Verlag der Wissenschaften, Berlin.

WAGNER, E. G., EHRIG, H. (1987) Canonical Constraints for Parameterized Data Types. *Theoretical Computer Science (to appear).*

WAND, M. (1979) Final algebra semantics and data type extensions. *Journal Computer System Science* **19**, 27—44.

WIRTH, N. (1971) Program development by stepwise refinement. *Comm. of the ACM* **14**, 4, 221—227.

ZILLES, S. N. (1980) Introduction to data algebras. In: *Proc. of Advanced Course on Abstract Software Specifications,* Copenhagen, LNCS 86, Springer-Verlag, Berlin—Heidelberg—New York, 248—272.

ZILLES, S. N., LUCAS, P., THATCHER, J. W. (1982) *A look at algebraic specifications.* IBM Res. Rep. RJ-3568.

Index